SpringerBriefs in Applied Sciences and Technology

SpringerBriefs present concise summaries of cutting-edge research and practical applications across a wide spectrum of fields. Featuring compact volumes of 50–125 pages, the series covers a range of content from professional to academic.

Typical publications can be:

- A timely report of state-of-the art methods
- An introduction to or a manual for the application of mathematical or computer techniques
- A bridge between new research results, as published in journal articles
- A snapshot of a hot or emerging topic
- An in-depth case study
- A presentation of core concepts that students must understand in order to make independent contributions

SpringerBriefs are characterized by fast, global electronic dissemination, standard publishing contracts, standardized manuscript preparation and formatting guidelines, and expedited production schedules.

On the one hand, **SpringerBriefs in Applied Sciences and Technology** are devoted to the publication of fundamentals and applications within the different classical engineering disciplines as well as in interdisciplinary fields that recently emerged between these areas. On the other hand, as the boundary separating fundamental research and applied technology is more and more dissolving, this series is particularly open to trans-disciplinary topics between fundamental science and engineering.

Indexed by EI-Compendex, SCOPUS and Springerlink.

More information about this series at http://www.springer.com/series/8884

Taehun Seung

Self-Induced Fault of a Hydraulic Servo Valve

A Possible Cause for Hidden Malfunction of Aircraft's Systems

 Springer

Taehun Seung
Ockenfels, Germany

ISSN 2191-530X   ISSN 2191-5318   (electronic)
SpringerBriefs in Applied Sciences and Technology
ISBN 978-3-030-03522-8   ISBN 978-3-030-03523-5   (eBook)
https://doi.org/10.1007/978-3-030-03523-5

Library of Congress Control Number: 2018962137

This Springer imprint is published by the registered company Springer Nature Switzerland AG
The registered company address is: Gewerbestrasse 11, 6330 Cham, Switzerland

Acknowledgements

The author would like to thank his colleagues at his former employer Mr. Thomas Fricker for the valuable contribution whilst troubleshooting and Mr. Klaus Baldauf and Mr. Rudolf Scheiblich for their indefatigable supports whilst laboratory investigation.

The same is valid for his colleagues at the partner companies in USA and Brazil.

The professional advice and proof-reading of Ms. Jane Lawson in writing this publication are also gratefully acknowledged.

Contents

1 Occurrence and Suspicion 1
  1.1 Occasion for Investigation, Brief Description
       of the Occurrence 1
  1.2 Appraisal and Fundamental Contemplation for a Pressure-
       Controlling Servo Valve 2
       1.2.1 Design Shape/Working Principle of the Hydraulic
              Valve 2
       1.2.2 Characteristic Diagram—Static and Dynamic Demanding
              and the Signal Answer 3
       1.2.3 Significant Influence Factors and Fault Sources of a Servo
              Valve 6
              1.2.3.1 Particle Contamination of the Fluid
                       (—A Widespread Superstition) 6
              1.2.3.2 Robustness of the Torque Motor (First Stage) ... 7
       1.2.4 Fault Potentials and Parameters in the Spool and Sleeve
              Assembly (Second Stage) 7
              1.2.4.1 Dormant Fault Potentials Hidden in Design
                       Shapes 7
              1.2.4.2 Significant Parametric Circumstances Whilst
                       Operating 13
       1.2.5 Fault Potentials/Possible Circumstances in a Solenoid
              Valve (First Stage) 15
  1.3 Existence of a Self-induced Fault of a Two-Stage Servo
       Valve—Hypothesis 17

2 Survey and Supposition 19
  2.1 Experimental Investigation—Test Execution and Results 19
       2.1.1 Test Set-Up and Test Method 19
       2.1.2 Measurement Data Analysis and Interpretations 21
              2.1.2.1 Abnormality in the Pressure Response
                       and Interrelations 21

        2.1.2.2 Stability of the Pressure Signal Answer and
                 Mobility of the Spool   21
        2.1.2.3 Switching-Off Dynamic of the Spool   24
        2.1.2.4 Feedback Intensity in the Water Hammer
                 Effect   31
        2.1.2.5 Fluid Dumping Behaviour at the Control
                 Circuit   32
        2.1.2.6 Hegemony Loss of the First Stage, Completion
                 of the Self-induction   38
        2.1.2.7 Development of a New Drainage and the Loss
                 of Command Ability   39
  2.2 Reconstruction of the Entire Fault Working Mechanism—Interim
       Conclusion   40

3 Supplements and Inference   41
  3.1 Additional Reflections/Supplementation of Contemplation   41
        3.1.1 Basic Working Mechanism of a Spool and Sleeve
               Assembly/Homologous Model   41
               3.1.1.1 Mobility of a Spool/Equilibrium at a Working
                        Point   41
               3.1.1.2 Confinement of the Spool's Freedom
                        Grade/Guided Sliding of the Spool   42
               3.1.1.3 Equilibrium and Meta-Equilibrium
                        in the Second Stage   43
        3.1.2 Origin of the Fault/Interlock Mechanism   48
               3.1.2.1 Order of the Fault Induction/Origin
                        of the Fault   48
               3.1.2.2 Self-stabilization of the Flapper/Interlock of the
                        Entire Servo Valve   49
               3.1.2.3 Danger of the Flow Separation on the Flapper
                        Body and the Timing   49
        3.1.3 Initializing, Propagation and Escalation of the
               Blemish/Reason for Blemish   52
               3.1.3.1 Degradation/Collapsing of the Lubrication
                        Film—'Dry Friction'   52
               3.1.3.2 Holistic Consideration Incl. The Reaction of the
                        Control Loop   53
        3.1.4 Arbitrariness of the Fault—Russian Roulette Effect   54
  3.2 Conclusion and Summary   56
        3.2.1 Conclusion   56
        3.2.2 Summary   58

Epilog   59

References   63

Abbreviations

A_C	Circular surface area—control side in the second stage
A_S	Circular surface area—coil spring side in the second stage
A/C	Aircraft
B	Brake port (hydraulic line)
BCV	Brake control valve
C	Control port (hydraulic line)
COMD	Commanded
GRB	Gear retract braking
F_H	Hydraulic force
F_{F1}	Friction force whilst pressurizing
F_{F2}	Friction force whilst releasing
F_R	Resulting radial force on spool a/o flapper
F_S	Coil spring force
h_1	Gap height at a coaxial aligned spool relative to the sleeve
h_2	Gap height at a coaxial misalignment of the spool relative to the sleeve
ICAS	International Council of the Aeronautical Sciences
Lohm	Liquid Ohm
P	Pressure port (hydraulic line)
PPT	Pressure pulse test
PR	Pressure
P_C	Pressure on circular surface area—control side in the second stage
P_S	Pressure on circular surface area—coil spring side in the second stage
PT	Pressure transducer
PR > C	Pressure more than commanded
R	Return port (hydraulic line)
Re	Reynolds number
S_1	Opening surface area at a coaxially aligned spool and sleeve assembly
S_2	Opening surface area at a misaligned spool and sleeve assembly
WT	Wheel speed transducer

Introduction

An intermittent uncontrollability of hydraulic servo valves equipped with spool and sleeve is called 'hydraulic locking' in technical terminology. This malfunction differs from the mechanical blockage of the spool caused by single or multiple foreign materials or mechanical deformation of a member.

Scientists and engineers have concluded that the spool tends to stick due to increased friction whilst operating [2, 14]. It seems that this type of malfunction occurs more often than that caused by debris. In many fault cases, no debris has been found despite premature supposition.

Meanwhile, thanks to the systematic analysis of research studies made over the last couple of decades, the physical characteristics of such sticking are understood [2, 14]. The investigations, however, have been mostly focused on the spool and sleeve assembly, although this is only a part of a servo valve. In contrast to the so-called direct drive, the spool of such conventional two-stage servo valve 'swims' freely inside the sleeve. Its movement will be controlled by a control device and return spring. Should the secondary stage be considered alone, the traditional term of 'hydraulic locking' is a bit of a misnomer and overstates the situation, as the spool in the secondary stage sticks only temporarily. Whenever such disturbances occur, adequate control logic can adjust the spool position providing the primary stage is still responding. This will cause a lag in the final signal answer. In the case of a closed loop control system, however, such deviations attract no attention unless a predefined threshold value and/or time limit has been exceeded.

In contrast, a real unrecoverable 'hydraulic lock' occurs in a two-stage servo valve when the closed loop control system is no longer able to manage the second stage by means of the primary stage. This does not necessarily mean that the secondary stage is blocked by debris or fluid lock. It is rather a result of an interaction when both stages mutually influence each other and eventually freeze.

Due to this, the closed loop control is no longer able to manage the system. This interaction is more complex than 'sticking' of a spool and sleeve assembly.

This report[1] details the investigation and result made recently in the field of aircraft engineering and introduces the entire working mechanism of a self-induced fault of a servo valve.

---

[1]This is the final comprehensive description a shortened congress paper for 31st ICAS originates from. The paper 'ICAS 2018_0853' was presented in Belo Horizonte, Brazil, on 10 September 2018. cf. https://www.icas.org/ICAS_ARCHIVE/ICAS2018/data/papers/ICAS2018_0853_paper.pdf

Topics and Arrangement of This Report

The self-induced fault of a servo valve is a complex phenomenon, of which the existence and working mechanism could be substantiated only by intricate investigations.

Following this section, in Chap. 1, Sect. 1.1 the occasion for investigation will be briefly reported. Then, prior to describing the results of the investigation and its conclusion, some fundamental contemplation and observations will be discussed in Sect. 1.2, for better understanding whilst reading the later chapters. A part of these discussions will be made on the basis of some measurement data and experiences collected during the troubleshooting phase.

Then, in Sect. 1.3, a hypothesis will be derived prior to discussing results from the experiments.

In Chap. 2, Sect. 2.1.2, some measurement data from the troubleshooting will be analyzed regarding the coherences of physical effects discussed in Sect. 1.2. The results of the analysis function as proof/evidence of the hypothesis derived in Sect. 1.3. In Chap. 3, Sect. 3.1, some additional explanations will be given to complete the observation. This will show that the self-induced fault introduced in this report may occur at any hydraulic servo valves of similar working principle. Finally, the entire working mechanism as discovered will be described in Sect. 3.2.1 as the final conclusion.

These approach steps are necessary in order to make the complex, simultaneously occurring phenomena understandable in this limited exposition with its few pages. Section 3.2.2 summarizes the present investigation results. References are listed in the book backmatter.

Chapter 1
Occurrence and Suspicion

1.1 Occasion for Investigation, Brief Description of the Occurrence

The state-of-the-art brake system of a modern transport aircraft is actuated electro-hydraulically. Such a so-called brake-by-wire system of a regional aircraft, of which manufacturer and brake system provider wish not to be mentioned by names here, failed sporadically whilst operating. The occurrences affected seriously the dispatch ability of the aircraft as one of the duplex brake systems switches off arbitrarily by itself without any typical error patterns (Nevertheless, it must be said that the passengers and crews were endangered at no time thanks to the fail-safe architecture of the brake control system. Moreover, a sufficient safety margin was guaranteed by the separated emergency brake system at the time of the events.). Figure 1.1 depicts the principle of a 'brake-by-wire' control system inclusive of emergency brake circuit.

The fault message of the system reads as: 'PR MORE THAN BCM COMD'. According to the message, the feedback pressure level at the brake cylinder was abnormally higher than the intended set value of the brake control system. In such condition, the control logic switches off the brake control valve (BCV) at the corresponding brake in order to prevent possible damage incurred by overheating. Note that the brake system will be switched off pairwise, either inboard or outboard brake circuit, whenever a brake in the corresponding circuit is faulty (cf. Fig. 1.1).

As the faults scattered almost evenly at all wheels, the troubleshooting was focused on the brake control units. The BCV, which is a pressure-controlling, two-stage hydraulic servo valve, was eventually under suspicion, even if tens of thousands of valves of a very similar type from the manufacture were working properly in other aircraft model ranges.

© The Author(s), under exclusive license to Springer Nature Switzerland AG 2019
T. Seung, *Self-Induced Fault of a Hydraulic Servo Valve*,
SpringerBriefs in Applied Sciences and Technology,
https://doi.org/10.1007/978-3-030-03523-5_1

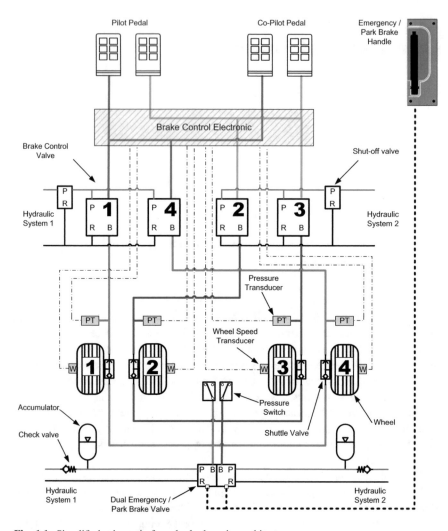

Fig. 1.1 Simplified schematic for a brake-by-wire architecture

1.2 Appraisal and Fundamental Contemplation for a Pressure-Controlling Servo Valve

1.2.1 Design Shape/Working Principle of the Hydraulic Valve

The brake control valve involved is a pressure-controlling, two-stage electromagnetic servo valve. This valve consists of a torque motor assembled with a metering valve of 'flapper and nozzle' type and a 'spool and sleeve' assembly accommodated

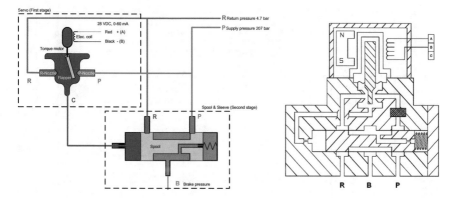

Fig. 1.2 Principle schematics of the servo valve

in a separate housing. The principle schematic is shown in Fig. 1.2. The primary stage employs no cantilever feedback spring, i.e. there is no mechanical corresponding connection between primary and secondary stages (cf. Sect. 3.2.2 Ref. [13]). A coil spring in the secondary stage resets the spool whenever the flapper reduces the control pressure to the secondary stage. It must be said that the reset mechanism has hardly any effect on the circumstances and phenomena described hereafter. Furthermore, the physical phenomena and resulting fault mechanism described below can occur in any kind of pilot-working hydraulic servo valves.

1.2.2   Characteristic Diagram—Static and Dynamic Demanding and the Signal Answer

Figure 1.3 shows the relation of 'pressure response versus spool stroke' in the secondary stage. The plots are made at a low, quasi-static demanding condition for which the spool was moving less than 0.16 mm/s.

Having passed a certain overlapped range at the beginning (i.e. 0–0.485 mm), the pressure answer corresponds directly proportional to the spool's stroke and consequently to the geometrical opening gap given between the edges of the spool and sleeve (i.e. 0.493–0.753 mm).

In contrast to the plots of quasi-static demanding shown above, Figs. 1.4, 1.5, 1.6 and 1.7 show some characteristic curves of the servo valve under different dynamic demanding conditions for which the demanding speed has been changed. It is easy to recognize that the hysteresis of the characteristic curve becomes more violent by increasing the demanding speed.

Regardless the demanding speed, i.e. despite quasi-static or dynamic demanding conditions, the spool adjusts at the actual working point to establish an equilibrium state, whereas the magnetic field strength and consequently the opening rate of the

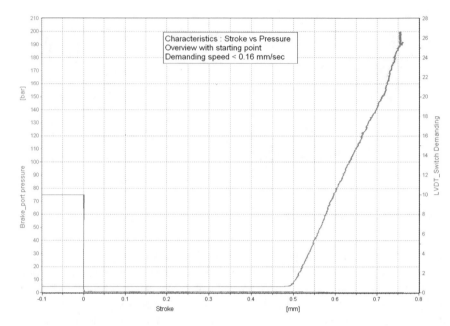

Fig. 1.3 Spool stroke versus Pressure response in the second stage

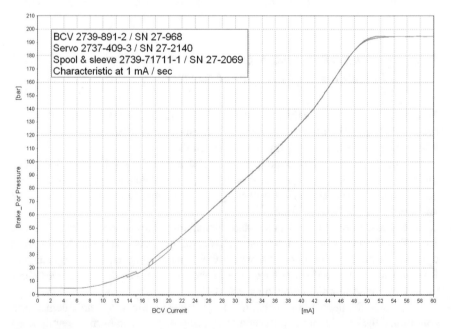

Fig. 1.4 Characteristic curve of the servo valve at a demanding speed of 1 mA/s

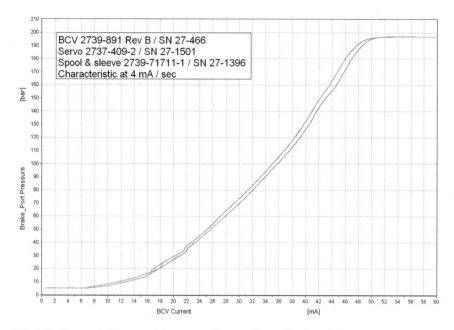

Fig. 1.5   Characteristic curve of the servo valve at a demanding speed of 4 mA/s

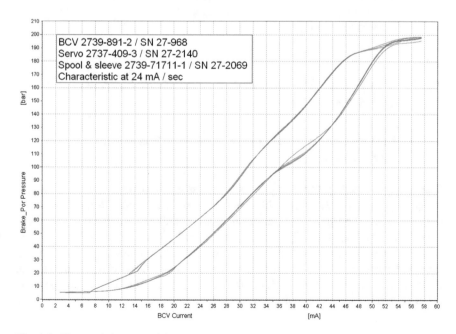

Fig. 1.6   Characteristic curve of the servo valve at a demanding speed of 24 mA/s

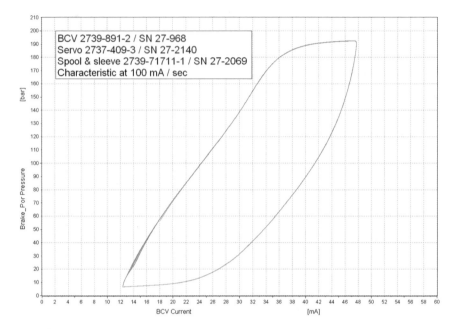

Fig. 1.7 Characteristic curve of the servo valve at a demanding speed of 100 mA/s

inlet nozzle in the first stage, friction/lubrication between spool and sleeve, internal leakage flow, the strength of the return spring, the inertia of the spool, etc., are the main parameters (cf. Sect. 3.1.1.1).

1.2.3 Significant Influence Factors and Fault Sources of a Servo Valve

1.2.3.1 Particle Contamination of the Fluid (—A Widespread Superstition)

Debris is prevalently taken as a possible root cause whenever a hydraulic valve is inexplicably faulty. It is conceivable that debris could influence the functionality of both first and second stages of a servo valve. It even militates in favour of the theory why some faults sporadically and only temporarily occur. Although the typical arbitrary characteristic sounds plausible in relation to the intermittent occurrences, such premature conclusion has often turned out to be too superficial and unfounded, as no debris was ever found in many fault cases. As a matter of fact, the possible effect of debris seems to be often overstated.

After having carried out extensive test campaigns with stepwise increased contamination rate, some experts have even concluded that the servo valves are very much more insensible than they are usually supposed to be [1]. A similar laboratory test campaign carried out for this actual investigation confirmed this. Moreover, despite sufficient filtration of the fluid, the occurrences mentioned in Sect. 1.1 were still recurring with nearly the same fault rate. So far, debris was excluded from the list of possible root causes for the present investigation.

As stated in the introduction, the present work deals solely with uncontrollability caused by purely fluid mechanical occurrences, i.e. non-debris-faults.

1.2.3.2 Robustness of the Torque Motor (First Stage)

In most cases, the torque motor consists of an electric coil and relatively simple (preloaded) mechanical moving part which stands under the effect of one or more permanent magnets. The electric coil can be a simplex or duplex type and creates/regulates the necessary magnetic field by demanding in order to control the initial movement of the (pilot-working) valve system.

Dropout of such electromagnetic parts due to any kind of interference, like foreign magnetic field, vibration/shaking and/or shock, is conceivable in spite of shielding/insulation. But this possibility is not to be taken into consideration already as a root cause since there was no typical error pattern. In the case of the present investigation, the fault still occurred at many serial units without changing in fault rate, despite extremely varying electromagnetic working conditions. The fault still occurred even in an insulated environmental condition (both vibration and magnetic field). Hence, it was uncertain that the fault had been caused by foreign influences —such as debris.

1.2.4 Fault Potentials and Parameters in the Spool and Sleeve Assembly (Second Stage)

1.2.4.1 Dormant Fault Potentials Hidden in Design Shapes

Dimensional Tolerances at the Spool and Sleeve Assembly

In terms of movement ability of a spool, its (radial) fit tolerance relative to the sleeve is of prime importance. It is even trivial to mention that the effective fit tolerance will be determined by the actual temperature. The development of modern numeric simulation tools has been made quite good progress during the last decades, and they are able to predict the possible thermal effects with a high level of accuracy. Thermal effects, therefore, can be considered as an underpart with minor importance, as long as this does not give rise to mutual effects with other parameters.

Besides the radial fit tolerance of a spool and sleeve assembly, the actual angular position of the spool relative to the sleeve could have a significant influence on the movement ability of the spool. This comes from the straightness tolerances of both spool and sleeve. In fact, neither the spool nor the sleeve is absolutely straight and free of bends since it is impossible to machine such a perfect geometrical shape. Should both members be in paraphrase with their bends at an indeterminate moment, i.e. convex to concave, the actual friction of the sliding surface would be maximized and vice versa (cf. Fig. 3.12). Note that this is only a reflection of a two-dimensional case as a simplified thought experiment. In reality, the curve progressions are three dimensional. In any case, the friction between the spool and sleeve changes in accordance with the actual angular position of the spool. Hence, the straightness of the members can be of decisive parameter and particular importance, even though this often is taken as a matter of course.

In order to investigate the possible influence of the spool's angular position on the pressure development in the characteristic diagram, a test campaign was made.

The spool was prepared with a mark so that different angular positions could be set relative to the sleeve (cf. Fig. 1.8. Note that the spool is actually set to the 12 o'clock position.). The test was performed with four different spool positions preset between the test campaigns.

Whilst field operating, the spool is able to roll inside the sleeve an indeterminate amount of angle and direction corresponding to the actual dynamic pressure, flow rate, leakage stream, running speed of the spool, etc. In order to keep the preset position of the spool, no consumer was connected to the valve. The flow rate, which is one of the main parameter, was eliminated during the whole test in this way. The test was conducted at a standardized procedure at which the running time and pause were kept same.

Fig. 1.8 Marking on the spool

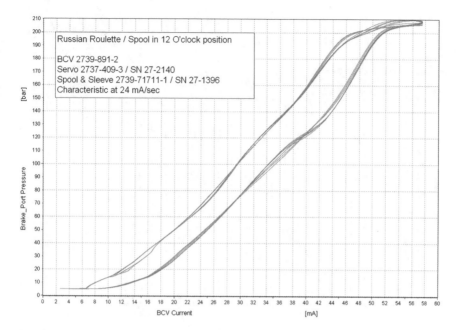

Fig. 1.9 Sensitivity in the 12 o'clock position

The measurement results plotted in Figs. 1.9, 1.10, 1.11 and 1.12 show clearly the significant influence of the spool's angular position on the pressure development (Every plot contains ten cycles demanded from 2.647 to 57.533 mA at a constant demanding speed of 24 mA/s.).

The curve runs shown in Figs. 1.9, 1.10 and 1.11 strew in a certain range and even change its gradient at each cycle. In contrast, the curve runs shown in Fig. 1.12 are strikingly smooth and keep its gradient throughout the whole series of cycles.

Comparing the diagrams with that in Fig. 1.12, it is concluded that the running speed of the spool changes and it is dependent on the actual angular position of the spool. Moreover, it is recognized that the sliding ability at the same point is getting worse with increasing number of cycles (cf. Figs. 1.9, 1.10 and 1.11). The reason for such changing in sliding ability seems to be the degeneration of the lubrication film caused by rubbing (cf. Sect. 3.1.3.1).

Note that Schlemmer et al. [12] carried out a similar investigation and reported a certain angle dependency of the running friction of the spool.

Importance of the Balancing Grooves for Spool's Working Ability

The balancing grooves on the spool are occasionally called 'relief grooves', and there are numerous detailed studies and suggestions in terms of form, number and

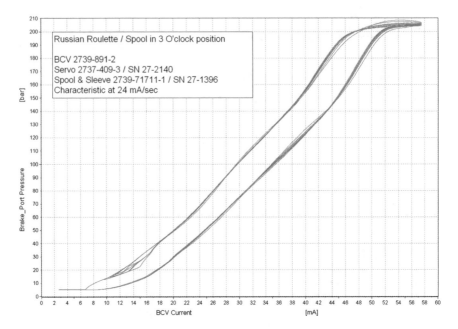

Fig. 1.10 Sensitivity in the 3 o'clock position

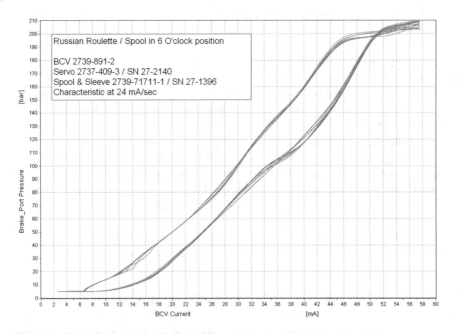

Fig. 1.11 Sensitivity in the 6 o'clock position

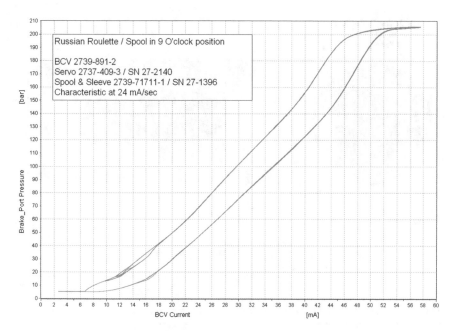

Fig. 1.12 Sensitivity in the 9 o'clock position

distance between each balancing groove [2, 3, 7–12, 14, 15]. In order to make the function and working mechanism of the balancing grooves clear, simplified qualitative contemplations will be briefly discussed: A balancing groove is able to work correctly only when the groove is able to build a closed annular chamber around the spool.

In other words, the functionality of balancing grooves may only be guaranteed when they are covered by the sleeve. Figure 1.13 shows the principle and functionality of such an annular pressure chamber; the fluid from the higher pressure side tries to compensate its pressure difference by leaking to the lower pressure side. During the process, the leakage stream will try to find a way of lowest resistance. Hence, the leakage stream takes the route with the largest geometric gap. This does not necessarily have to be a direct route. In the reality, the route can be slightly bent in a radial direction so that it looks like a spiral line. Without having a working balancing groove, the pressure will be decreased proportionally among the streamline (cf. Fig. 1.14). Doing that, the pressure force would press the spool to the sleeve wall. The gap between the spool and sleeve could be increased as the spool's centreline is no longer aligned coaxially with that of the sleeve. Once a small misalignment occurs, the situation will become worse and worse at every movement. The spool is not able to recover the misalignment as the side force will be radically increased.

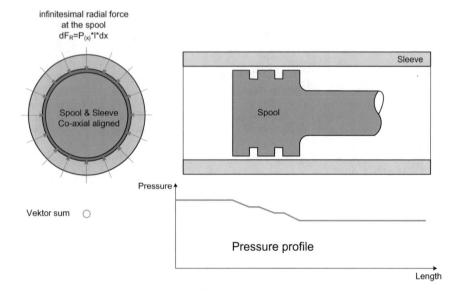

Fig. 1.13 Abolition of infinitesimal radial forces at perfect working balancing grooves as the result of the coaxial alignment of the spool relative to the sleeve

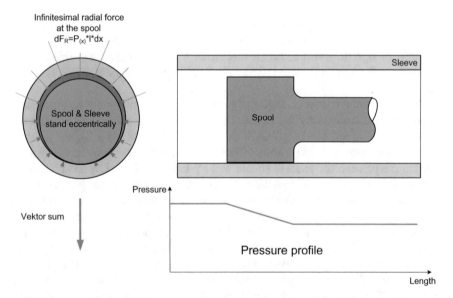

Fig. 1.14 Development of a resulting radial force at missing/non-working balancing grooves as the result of the misalignment of the spool relative to the sleeve

Should a balancing groove exist and still be working, the local pressure difference around the spool will be balanced since the local pressure around the spool's circumference changes thanks to the closed annual chamber equally and simultaneously [5]. During this process, the spool recovers a possible coaxial misalignment by itself (cf. Fig. 1.13). In other words, this means that the spool 'relieves' by itself. As a result, the leakage stream in circumference of the spool is evenly dispensed and consequently the flow rate, i.e. leakage, is timely constant (cf. Sect. 2.1.2).

1.2.4.2 Significant Parametric Circumstances Whilst Operating

This section summarizes some important signs/appearances and significant parameters which cause the spool and sleeve to stick. Although some parameters play dominant roles under certain conditions, they are often classified in the practices even as trivial with negligible effects. Hence, some issues to be easily overlooked will be highlighted here.

Changing of Actual Friction on the Sliding Surface Area Whilst Operating

Excluding debris and dropout of the electric parts as possible root causes (cf. Sects. 1.2.3.1 and 1.2.3.2), there are hardly any other fault-provoking parameters and/or problem potentials than friction on the sliding surface area of the spool and sleeve assembly.

This kind of mechanical problem known as 'hydraulic locking' has existed for ages, practically since the spool and sleeve assemblies came into use for hydraulic control devices.

As briefly mentioned in Sect. 1.1, this traditional terminology is a bit of a misnomer, in so far as it is used for the spool and sleeve assembly. It expresses the situation as if the valve were blocked fast and durably. In reality, the spool sticks only temporarily, so far no deformation has occurred at a member (i.e. either spool or sleeve) or a certain situation freezes the situation (cf. Sect. 1.3). Some parametric studies about such 'sticking' of a spool and sleeve assembly were made, and countermeasures have been developed [2–5, 7–12, 14, 15].

Internal Leakage and Gap Tolerance as a Significant Influence Factor

It is fully impracticable and undesirable to assemble the spool and sleeve completely without internal leakage unless the motion of the spool is managed by high actuation force, e.g. direct drive. Having an adequate amount of internal leakage, the spool is able to 'swim' inside sleeve. The leakage stream in the gap between the spool and sleeve flushes possible impurity, and above all the fluid functions as a lubricant. The prerequisite, however, is an even gap distance throughout the whole

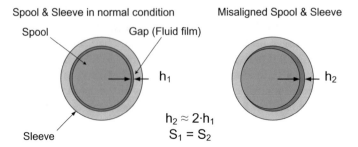

Fig. 1.15 Change of the gap distance despite same opening surface area

circumference of the circular gap (cf. Section "Importance of the Balancing Grooves for Spool's Working Ability").

Should the spool be placed eccentrically, the internal leakage and consequently the purging gap stream increases exponentially despite unchanged absolute opening surface area, since the leakage flow is not dependent on the surface opening rate but on the gap amount according to the two-dimensional Reynolds equation given below [9, 12]:

$$\frac{\partial}{\partial x}\left(\frac{h^3}{\eta}\cdot\frac{\partial p}{\partial x}\right)+\frac{\partial}{\partial z}\left(\frac{h^3}{\eta}\cdot\frac{\partial p}{\partial z}\right)=6\cdot\frac{\partial}{\partial x}\left(\frac{\partial(h\cdot U)}{\partial x}+\frac{\partial(h\cdot W)}{\partial z}+2\cdot\frac{\partial h}{\partial t}\right) \quad (1.1)$$

where

h: gap height, p: pressure, U: velocity in x-direction
x: tangential coordination, W: velocity in z-direction
z: tangential coordination, η: dynamic viscosity

Figure 1.15 shows the geometric condition in the case of a metallic contact due to a non-coaxial misalignment of the sleeve. On the opposite side of the contact point, the gap amount increases to double compared to the original gap distance and consequently the internal leakage, in other words the 'purging quantity', increases abruptly according to the equation given above. Once a continuous axial flow is established, the spool is hardly able to recover the original gap by itself because the resulting lateral force of the spool presses the spool to the sleeve wall (cf. Fig. 1.14).

In the case of a radial contact of the spool to the sleeve's wall, it is also to expect that the friction will be drastically increased.

Air/Temporary Vacuum in the Hydraulic Circuit

Air in the hydraulic circuit is not only as a result of an insufficient bleeding process but also accrues in fact as a 'generative' problem. The amount of air in the hydraulic fluid increases during the operation as the mechanical parts, which are shuttling two medium zones, like the piston rod, will continue to bring new gas molecules into the fluid side at every stroke.

Vacuum accelerates the 'sliding-surface degeneration' process: due to a variety of reasons, the pressure of a hydraulic circuit could sink down locally even below atmospheric pressure during the operation. In such a case, the gas molecules absorbed in the fluid can suddenly escape and create bubbles.

A detailed analysis was made during the troubleshooting phase of the current investigation. Once bubbles are created, significant 'sponge effect' was the consequent result at least in the following command cycle in the case of a brake-by-wire control system.

1.2.5 Fault Potentials/Possible Circumstances in a Solenoid Valve (First Stage)

Generally, the first stage is less sensitive than the second stage even if it has more parts and requires a fine adjustment in most cases. Once assembled properly and protected by an adequate 'last chance filter', shielded against electromagnetic effects and protected against possible vibrations, the first stage is rather robust. This, however, does not necessarily mean that the flow control mechanism of the first stage is immune to any kind of instability. The system, for example, can react sensitively to fluid mechanical phenomena: should the control system accidentally work across a working point, at which the flow separation occurs, the effect of such a transition must be noticeable in the electro-hydraulic signal conversion (unique incidence working point).

By reaching the sub-critical flow region ($10^3 < Re < 1.7 \sim 4 \times 10^5$), the flow separates on the cylindrical flapper body, so that this experiences an extra drag. The flapper drifts away to the downstream direction as if a sudden suction force would have been developed on its lee side. Whenever such flow separation occurs on the flapper body, the gap at the nozzle increases accordingly a certain amount due to the extra force.

The effect of such an extra opening at the nozzle clearly reflects in the pressure answer as shown in Fig. 1.16 as a sudden increase in the gradient (see Sect. 2.1.1 for the pressure sensor positions). The extra opening of the nozzle will be kept until the reversal of the command—as long as the system is able to create a sufficient reset force afterwards to manage the flapper movement (cf. Sect. 3.1.2).

It must be said that such a working point is difficult to detect in the reality due to numerous, simultaneously changing nonlinear parameters, even though there is no doubt of its existence.

Figure 1.17 shows the corresponding time history of Fig. 1.16. It is easy to recognize that the pressure answer sporadically inclines to override at a preset current limitation of 19.8 (mA).

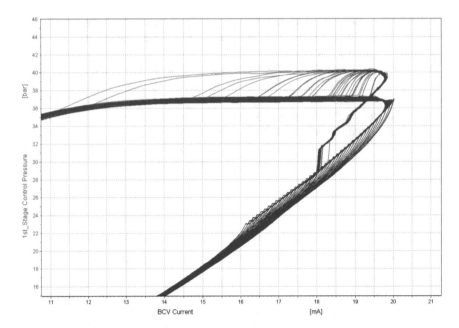

Fig. 1.16 Sporadic sudden increase of the pressure due to extra suction force at the flapper

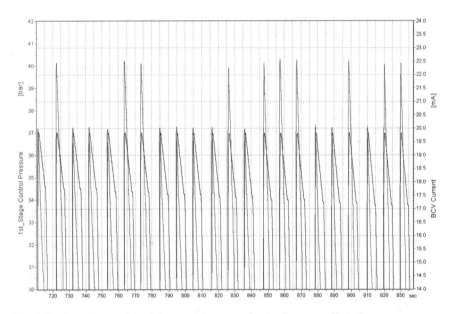

Fig. 1.17 Sporadic override of the pressure answer despite the current limitation

Fig. 1.18 Principle schematic—development of a relative force (drag) in the downstream direction due to the flow separation on the cylindrical flapper body

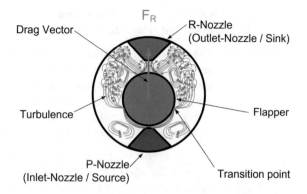

Figure 1.18 shows the principle of an extra force development at a cylindrical flapper body.

Whenever this happens, the second stage follows the signal from the first stage accordingly so that a discontinuity in gradient is to be found in the curve of the characteristic diagram (cf. Fig. 3.7. Note that this still happens only in a unidirectional manner keeping the hierarchical order given between the first and second stages, i.e. there is no feedback from the second stage to the first stage.) As long as the control system passes such a working point quickly enough, the discontinuity has no significant effects on the signal answer and consequently does not seriously affect the operation. On the other hand, the present investigation showed that such flow separation at a cylindrical flapper body can introduce a self-induced fault of a two-stage servo valve if the system dwells at the working point. However, it does not necessarily mean that the first stage is the origin of the self-induced fault as the phenomenon at the flapper only amplifies the pressure signal answer in the first place (cf. Sect. 3.1.2).

Besides specific fluid dynamic occurrences and static internal leakage, the other possible disturbances discussed in the previous chapters, such as debris and trapped air bubbles, can certainly affect a solenoid valve during its operation.

1.3 Existence of a Self-induced Fault of a Two-Stage Servo Valve—Hypothesis

During the troubleshooting phase, it was ascertained that there must be a more complex fault with mutual influences between the first and second stages besides the simple fault caused by sticking spool in the second stage (cf. Sect. 1.2.4.1). It seems to be a more complex fault with a self-induction, whereas the entire system is simultaneously influenced by numerous nonlinear parameters. In the worst case, the whole valve system can even be completely 'frozen'.

Because it is rather a complex phenomenon, a hypothesis seems to be useful and appropriate to relate and categorize the measurement results:

Hypothesis: The two-stage hydraulic servo valve built according to the principle design shape shown in Fig. 1.2 has a certain singular operational range, at which the hydro-mechanical phenomena occur intermittently in the second stage, and in such a strength that the first stage induces mutual interactive effects so seriously that the system eventually and irrevocably loses its controllability.

In the next chapters, the existence of such a complex phenomenon will be evidenced step by step by means of some fluid mechanical observations/analysis based on measurement data gained from both aircraft and laboratory tests.

Chapter 2
Survey and Supposition

2.1 Experimental Investigation—Test Execution and Results

2.1.1 Test Set-Up and Test Method

Figure 2.1 shows the principle schematic of the test set-up. This reflects the original A/C flight test instrumentation with which the hydraulic pressure levels are monitored for the supply and return lines as well as for the outlet port of the brake control valve. As the attention was focused on the pressure development and its variation at whole system members, an extra pressure transducer was installed between the first and second stages. Data acquisition was conducted at a sampling rate of 200 Hz.

The major difficulty during the troubleshooting was to find the exact working point of the sporadically failing pressure control system (brake-by-wire system). Although the typical fault condition was more or less known already, an exact determination of the working point was not a trivial issue due to the numerous, continuously changing nonlinear parameters.

Hence, the test was conducted in such a way that a certain command profile was repeated in 'endless' manner, and then the pressure, temperature, etc., were systematically changed. The command profile consists of a self-test impulse and a terminating command, with which the system performs an in situ system-check prior to lowering the landing gear in the approach flight phase and decelerates the rest spin of the wheels before retracting the landing gear into the bay after take-off.

The test cycle was repeated until a series of faults was eventually registered. Figure 2.2 presents therefore only a small section of the entire test measurement data. The section recorded for approximately 160 s contains six system-check impulses called 'pressure pulse test', and the same number of so-called 'gear retract braking' ramps in change. Note that the valve was switched off at every single impulse or ramp.

© The Author(s), under exclusive license to Springer Nature Switzerland AG 2019
T. Seung, *Self-Induced Fault of a Hydraulic Servo Valve*,
SpringerBriefs in Applied Sciences and Technology,
https://doi.org/10.1007/978-3-030-03523-5_2

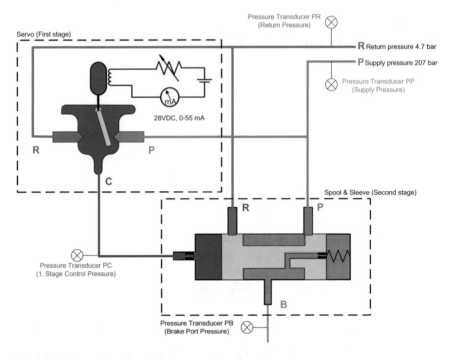

Fig. 2.1 Test set-up/position of the pressure transducers

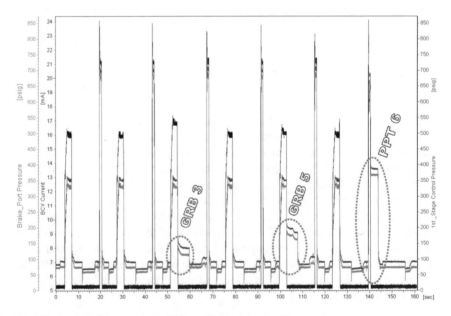

Fig. 2.2 A selected test result for 160 s with three abnormalities

For better understanding, the discussions in the following chapters refer to the 'gear retract braking (GRB)' and 'pressure pulse test (PPT)' with their event numbers. For example, 'GRB 1' means the first terminating command started at $t = 4$ s, whilst 'PPT 6' means the sixth in situ system-check carried out at $t = 139.4$ s., etc.

2.1.2 Measurement Data Analysis and Interpretations

2.1.2.1 Abnormality in the Pressure Response and Interrelations

As described in Sect. 1.1, the fault becomes noticeable due to abnormal higher pressure, i.e. 'pressure more than commanded'. Figure 2.2 shows three abnormalities as such. It must be said that the classification for abnormality is dependent on the threshold values of the pressure monitoring in the control loop (in most cases both pressure level and dwell). The first two abnormalities in the record would not be recognized as faults in A/C due to their low pressure level and relative short dwell. In any case, these three abnormalities contributed valuable realizations to the troubleshooting.

Prior to discussing the measurement results in detail, some interrelation facts will be clarified:

- The pressure response at the brake port is determined by the geometric spool position and dependent on its actual running speed (cf. Sect. 1.2.2).
- The gradient in the brake-pressure curve reflects the actual running speed of the spool.
- As 'communicating vessels', the ports and their associated components have mutual influences.
- The power consumption of the torque motor and consequently the current measured at the magnetic coil depend on the actual load applied to the hydro-mechanical part of the first stage of the valve.

2.1.2.2 Stability of the Pressure Signal Answer and Mobility of the Spool

Comparing the fault cases with normal cycles, it is striking that there are significant differences in pressure variations at the brake port (cf. Figs. 2.3, 2.4, 2.5, 2.6 and 2.7).

Considering that the pressure variation at a certain pressure level reflects nothing but a small dithering of the spool in its actual position, it seems that the spool must be kept in motion in order not to fail: in all three abnormal cases, the motion of the spool becomes 'hyper-stable' before the 'PR > C' event occurs. As long as the spool remains 'nervous', the valve works fine. There are even cases at which the

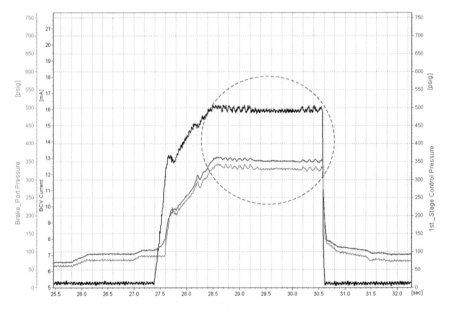

Fig. 2.3 Dithering recovery during the GRB No. 2; normal/no fault

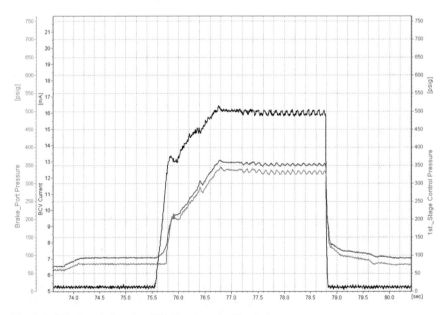

Fig. 2.4 Dithering during the GRB No. 4; normal/no fault

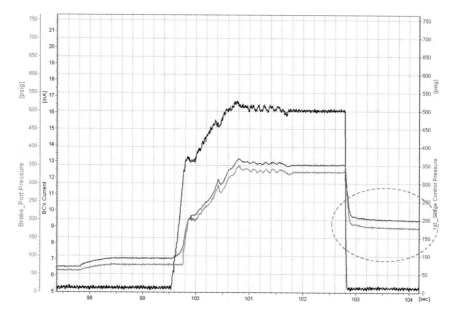

Fig. 2.5 Stop dithering during GRB No. 5; abnormal

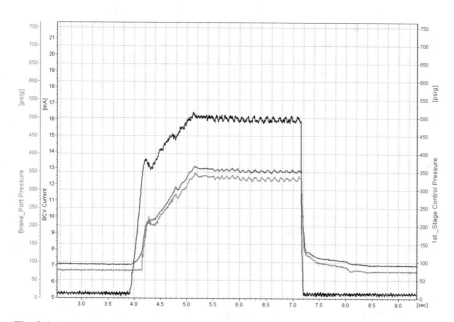

Fig. 2.6 Dithering during GRB No. 1; normal/no fault

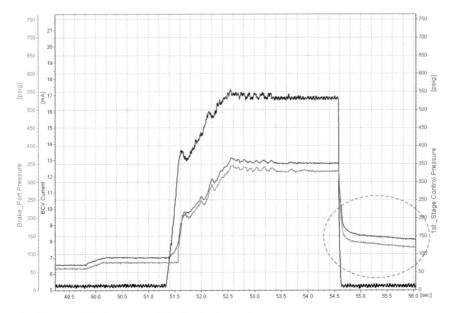

Fig. 2.7 Stop dithering during GRB No. 3; abnormal

spool recovers its dithering after having been 'hyper-stable' (cf. Fig 2.3, GRB 2). The system was not faulty in such cases.

It must be said that the mobility of the spool at a given command level depends mainly on the friction (cf. Sects. 2.1.2.3 and 3.1.1).

This is the first direct indication that the fault must be a mechanical problem or at least initialized by one or more mechanical parametric disturbances (Figs. 2.8, 2.9, 2.10 and 2.11).

2.1.2.3   Switching-Off Dynamic of the Spool

Considering that the viscosity and the bulk modulus are constant within a short time interval, the switching-off dynamic of the spool must be quite similar whenever the valve is switched off at a more or less equal spool position. Such changes in characteristic diagrams, however, are only recognizable when an adequate scale is chosen; cf. Fig. 2.12 versus Figs. 2.13 and 2.14.

Figures 2.14, 2.16 and 2.18 show a significant difference in the gradient of their curves compared to those of corresponding normal cases shown in Figs. 2.13, 2.15 and 2.17, whereas Figs. 2.19 and 2.20 compare the switching-off dynamic between PPT 3 and GRB 6 in the same axis scales. Note that the gradients of the brake-pressure curve at the switching-off phase are different in the cases of PPT and GRB because of the different operating loads of the bias spring.

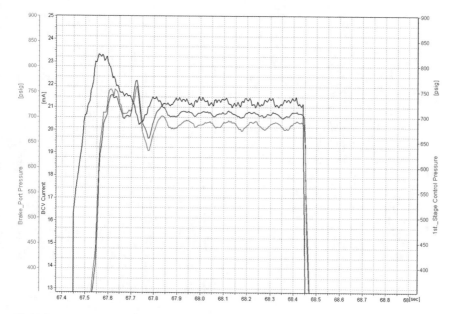

Fig. 2.8   Dithering during PPT No. 3; normal/no fault

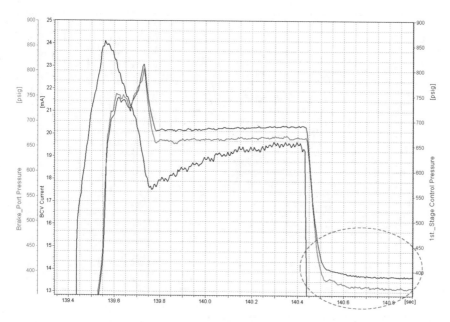

Fig. 2.9   No dithering during PPT No. 6; abnormal/fault

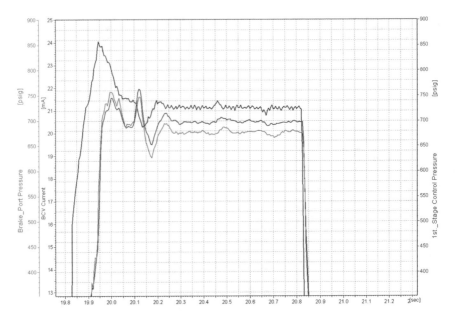

Fig. 2.10 Dithering during PPT No. 1; normal/no fault

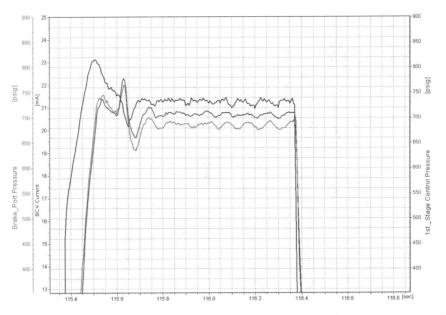

Fig. 2.11 Dithering during PPT No. 5; normal/no fault

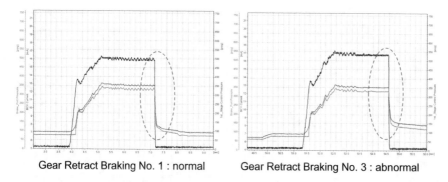

Gear Retract Braking No. 1 : normal   Gear Retract Braking No. 3 : abnormal

Fig. 2.12 Comparison of changing characteristics in two switching-off cases

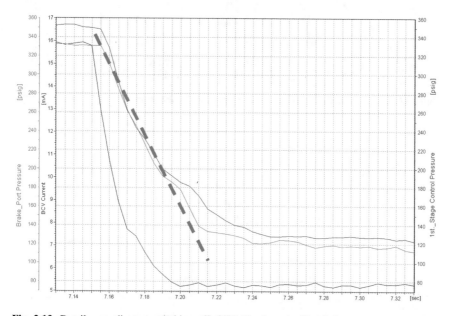

Fig. 2.13 Details—gradient at switching-off: GRB No. 1; normal/no fault

In the case of normal operation, the signal response at the 'switch off', i.e. the pressure dumping at the brake port, is rapid and similar under the same command type. Despite the very same switch-off condition, the pressure dumping process shows significant differences in the fault cases. The movement of the spool seems to become more sluggish since the gradient of the curve reflects in the first instance the running speed of the spool (cf. Section "Dimensional Tolerances at the Spool and Sleeve Assembly"). Again, the reason for changing in the running speed of the spool can only be the change of the actual friction between the spool and the sleeve.

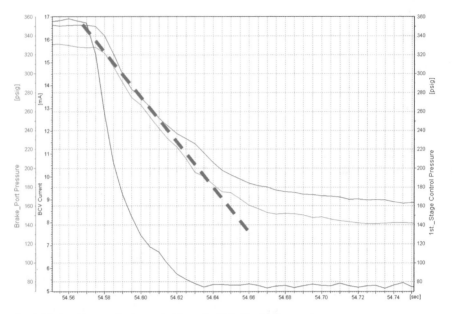

Fig. 2.14 Details—gradient at switching-off: GRB No. 3; abnormal

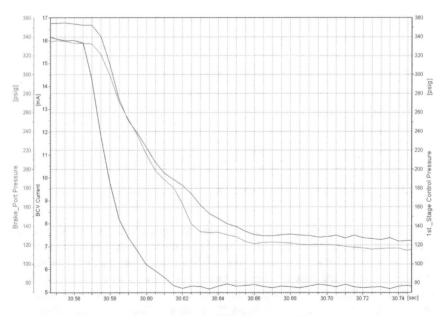

Fig. 2.15 Details—gradient at switching-off: GRB No. 2; normal/no fault

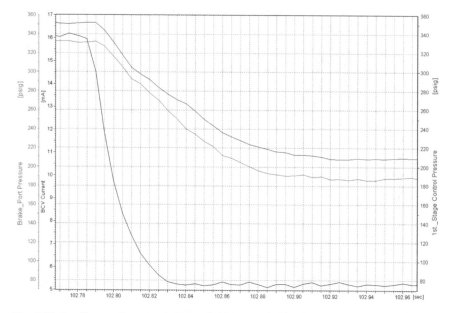

Fig. 2.16 Details—gradient at switching-off: GRB No. 5; abnormal

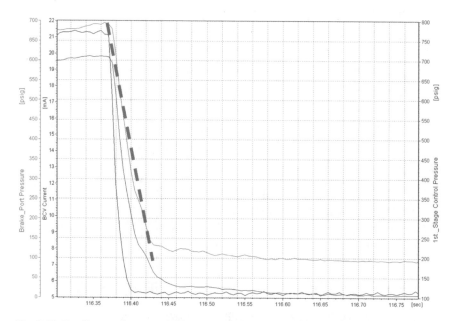

Fig. 2.17 Details—gradient at switching-off: PPT No. 5; normal/no fault

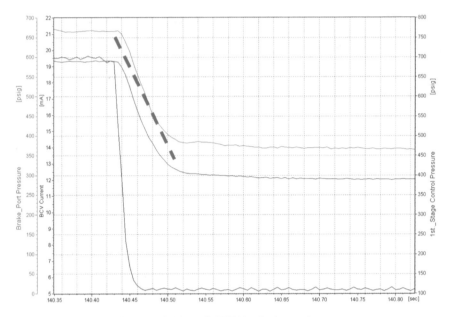

Fig. 2.18 Details—gradient at switching-off: PPT No. 6; abnormal

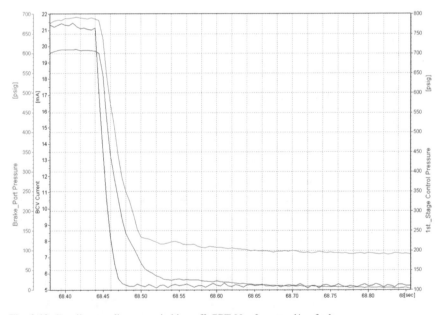

Fig. 2.19 Details—gradient at switching-off: PPT No. 3; normal/no fault

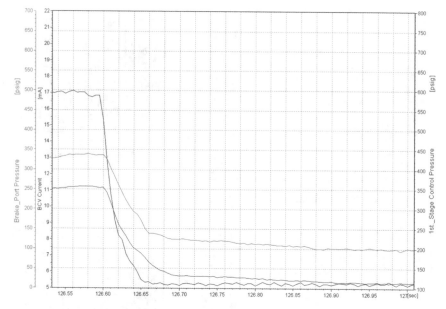

Fig. 2.20 Details—gradient at switching-off: GRB No. 6; normal/no fault

2.1.2.4 Feedback Intensity in the Water Hammer Effect

The feedback pressure level at the return line offers an extra valuable indication to corroborate the interpretation of friction changing in the second stage. The pressure impulse caused by a radical conversion of kinetic energy into static pressure, known as 'water hammer effect', shows a significant difference in the fault cases: in the case of a normal operation, a proper water hammer effect is to be found. In the fault cases, however, it is far less intensive due to the unintended but gentle closing of the brake port (cf. peaks of the return pressure listed in Tables 2.1 and 2.2 in Fig. 2.21).

A discrete changing from 'radical and abrupt' to 'slow and gentle' manner in closing of the port is to be concluded as a result of changed friction between the spool and the sleeve as long as the reset of the spool is managed only by a bias spring.

Figures 2.22, 2.23, 2.24, 2.25, 2.26 and 2.27 show pairwise the clear differences of the return pressure peak between the normal and fault cases after the switching off the electric coil.

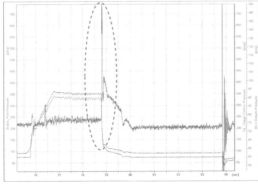

	RPmax [psig]	at t = [sec]
GRB1	147.55	7.180
GRB2	**150.83**	**30.595**
GRB3	137.40	54.600
GRB4	148.17	78.810
GRB5	139.12	102.820
GRB6	146.77	126.625

Tab. 6-1 Return Pressure
Peaks
at the end of the GRB

Gear Retract Braking No. 2 : normal

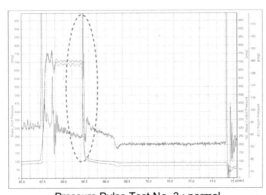

	RPmax [psig]	at t = [sec]
PPT1	215.78	20.845
PPT2	210.16	44.410
PPT3	**219.06**	**68.465**
PPT4	200.48	92.425
PPT5	205.48	116.390
PPT6	125.38	140.460

Tab. 6-2 Return Pressure
Peaks
at the end of the PPT

Pressure Pulse Test No. 3 : normal

Fig. 2.21 A typical water hammer effect at the switching-off (pressure peak in the return line)

2.1.2.5 Fluid Dumping Behaviour at the Control Circuit

In normal cases, whenever the solenoid is switched off and accordingly the flapper closes the inlet nozzle due to the spring preload, then the control circuit dumps the fluid trapped in the spool's control chamber via the outlet nozzle as the coil spring resets the spool back to the start position (cf. Fig. 1.2).

Such dumping processes are traceable in the record of the return pressure. For better understanding, the following description refers to the Figs. 2.28 and 2.29, of which time frames are from 19.0 to 24.0 s, from 139.0 to 143.3 s, respectively. It must be said that the system behaviour is generally quite reproducible, regardless of the ramp profile of the commands (cf. Fig. 2.30 and 2.31).

The dumping process seems to take 800 ± 100 (ms) depending on the last pressure level in the control chamber. Completing the process, the return line pressure stabilizes to a slightly lower level compared to that adjusted once under the operational condition (approximately 78 psi).

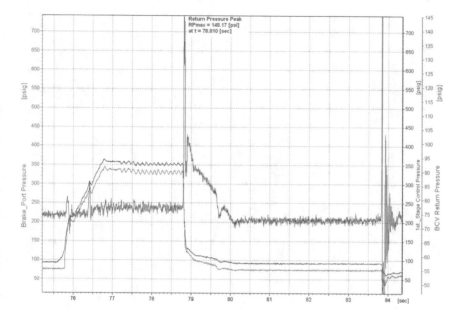

Fig. 2.22 Full water hammer effect at the switching-off: GRB No. 4; normal/no fault

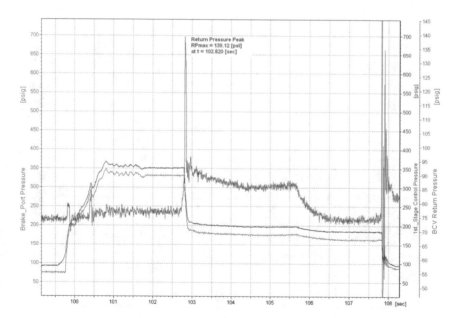

Fig. 2.23 Reduced water hammer effect at the switching-off: GRB No. 5; abnormal

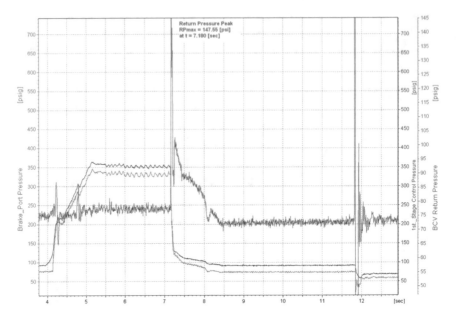

Fig. 2.24 Full water hammer effect at the switching-off: GRB No. 1; normal/no fault

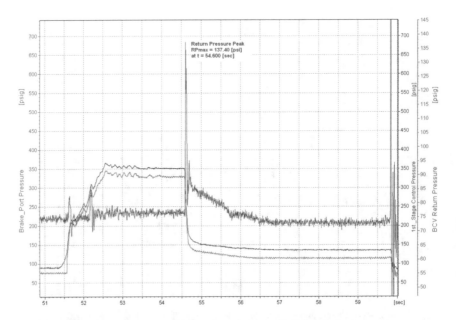

Fig. 2.25 Reduced water hammer effect at the switching-off: GRB No. 3; abnormal

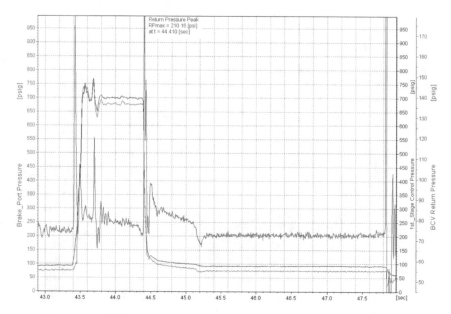

Fig. 2.26 Full water hammer effect at the switching-off: PPT No. 2; normal/no fault

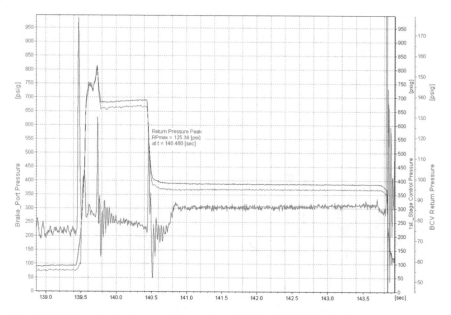

Fig. 2.27 Reduced water hammer effect at the switching-off: PPT No. 6; abnormal/fault

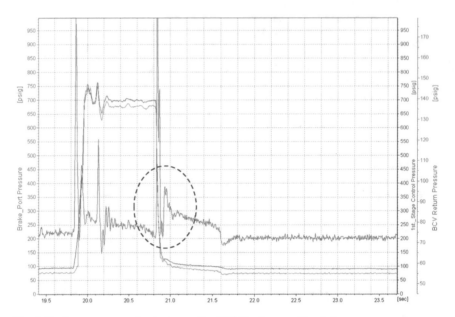

Fig. 2.28 Return pressure development after PPT No. 1; normal/no fault

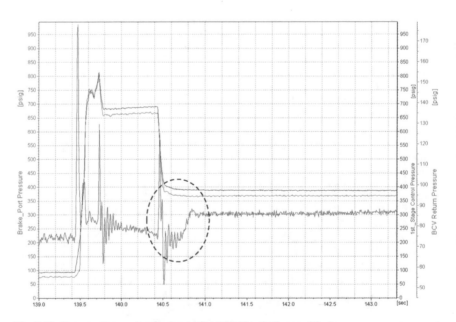

Fig. 2.29 Return pressure development after PPT No. 6; abnormal/fault

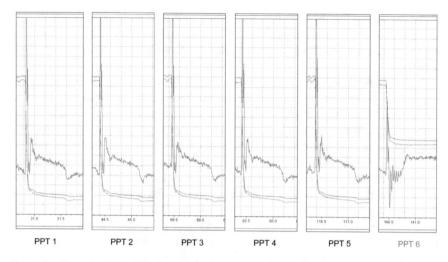

Fig. 2.30 Reproducibility of the pressure behaviour during the pressure pulse test and significant difference in the case of the fault

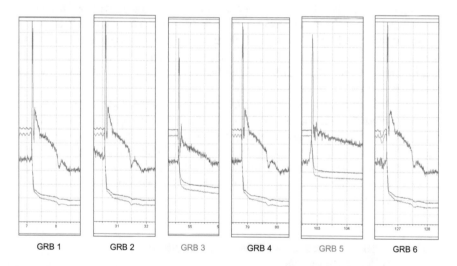

Fig. 2.31 Reproducibility of the pressure behaviour during the gear retract braking and significant difference in the case of the fault

Note that the outlet nozzle drains continuously a small amount of fluid whilst the solenoid is demanded. This is not leakage but 'regular fluid consumption'. This is the reason why the return pressure level sinks slightly below when the flapper is completely closed (72 psi at $t \geq 21.8$ s/Fig. 2.28).

In the case of fault, however, the return pressure stabilizes in an increased pressure level (approximately 86 psi at $t \geq$ 140.9 s/Fig. 2.29). This is definitely a sign that the inlet nozzle is not yet completely closed in spite of the switched-off solenoid.

It is striking also that the return pressure sinks down right after the switching off temporarily to the lowest possible level (for the present case approximately 75 psi at $t \approx$ 140.6 s/Fig. 2.29) before the return pressure increases to approximately 86 psi and stabilizes there. The temporal reduction of the return pressure means that the flow resistance (i.e. Lohm value [6]) between the pressure transducer PR and the inlet nozzle, i.e. downwards from the source (cf. Fig. 2.1), has been significantly reduced. The only possible reason for such a sudden reduction of the flow resistance is an extra drain path due to a misalignment of the spool in the sleeve (i.e. non-coaxial alignment of the spool; cf. Section "Internal Leakage and Gap Tolerance as a Significant Influence Factor"). In the normal case shown in Fig. 2.28, the water hammer effect disturbs the pressure dumping process. The bias spring in the second stage overcomes the disturbance and continues the process. The changing in the curve's gradient at $t = 21.06$ s reflects this. The buckle, however, is only a result of the superposition of the pressure wave created by the water hammer effect and the dumping pressure created by the bias spring. A damping effect against the direction changing of the pressure wave is recognized at $t = 21.03$ s. The dumping process seems to be continued until $t = 21.63$ s.

No matter what kind of mutual interferences ever happen between the first and second stages, the first stage has in the normal case only one drain, which is the regular outlet nozzle (R-nozzle). And the flapper is able to close reliably the inlet nozzle (P-nozzle).

In the fault case shown in Fig. 2.29, however, the flapper seems not to have a chance to close the inlet nozzle. Taking a new extra drain route, i.e. an additional connection to the return line via the gap between the spool and sleeve wall (cf. Section "Internal Leakage and Gap Tolerance as a Significant Influence Factor"), the inlet nozzle might have developed a constant internal flow as a result of the dramatically increased internal leakage in the spool and sleeve assembly (cf. t 140.9 s/Fig. 2.29).

2.1.2.6 Hegemony Loss of the First Stage, Completion of the Self-induction

According to the plots shown in Fig. 2.29, the return pressure sinks down right after the switching off to approx. 75 psi before it increases again to approx. 86 psi and stabilizes there. As discussed in Sect. 2.1.2.5, this is the evidence that the inlet nozzle in the first stage is still open because the return pressure does not sink down further to approx. 72 psi level. Hence, an internal flow must have been developed strong enough across the first and second stages to keep the inlet nozzle in an open position. In other words, the suction force at the flapper is so high that the flapper is no longer able to close the inlet nozzle by itself. It seems that the flow around the flapper separates the cylindrical flapper body during the process ($140.7 \leq t$ 140.9) by which the flapper experiences an extra amount of suction force (cf. Sect. 1.2.5). Having an extra amount of force, the resulting vector sum might be strong enough to overcome the closing spring force of the flapper or even to demand a further opening. The first stage loses definitively the hegemony in any case due to the fully developed, steady turbulent flow around the flapper (cf. Sect. 3.1.2.2).

2.1.2.7 Development of a New Drainage and the Loss of Command Ability

Considering the coherency between the internal leakage and effective gap distance discussed in Section "Internal Leakage and Gap Tolerance as a Significant Influence Factor", it is clear that the first stage drains in the case of spool's misalignment its incoming fluid suddenly in two routes. Figure 2.32 depicts the situation schematically. The channel between the first stage and the second stage, i.e. the route No. 2 in the figure, is no longer a bidirectional command line. This channel now works parallel to the original outlet channel of the first stage, i.e. the route No. 1, as a new unidirectional drainage. The severe implication is that the flapper is no longer manageable due to the high flow force at the inlet nozzle.

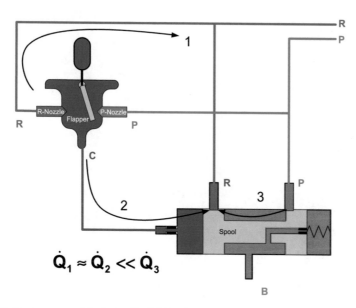

Fig. 2.32 Drain routes of the hydraulic fluid in the case of the fault

2.2   Reconstruction of the Entire Fault Working Mechanism—Interim Conclusion

Implications based on the physical phenomena discussed in Sect. 1.2, as well as the data analysis and interpretations described in Sect. 2.1, allow a reconstruction of what the root cause of a self-induced hydraulic locking looks like:

When the spool in the second stage of a hydraulic servo valve becomes sluggish due to increased internal friction, the orientation of the spool could be occasionally aligned no longer coaxial relative to the sleeve. If such a misalignment occurs, the effective gap distance between the spool and sleeve can increase up to 100%. Accordingly, the internal leakage increases abruptly and a constant flow develops in axial direction among the spool and sleeve gap. Then, the first stage dumps no longer solely via the outlet nozzle but also via the communicating vessel, i.e. through the control line additionally into the return line. The flow rate at the inlet nozzle increases dramatically since the sink pressure at the opposite side has been decreased. The arising effect is that the second stage seriously affects the controllability of the first stage as the flapper is hardly able to close the inlet nozzle. At a certain working point where the flow separating point on the flapper's lee side wanders in an upstream direction, the increased flow rate amplifies therefore the resulting flow force. Hence, the opening rate at the inlet nozzle increases. This again increases the total internal leakage/drain at the second stage. Once such a mutual influence is initialized, the escalation occurs so rapidly that the system does not have any chance to recover its controllability and eventually freezes.

Chapter 3
Supplements and Inference

3.1 Additional Reflections/Supplementation of Contemplation

Some details of the working mechanism and additional considerations will be discussed in this section. The discussion will point out some essential phenomena which occur in a spool and sleeve assembly during operation and in the case of a fault. The present description is valid for both pressure control and flow control valve systems in so far as they employ a pilot-working spool and sleeve assembly.

3.1.1 Basic Working Mechanism of a Spool and Sleeve Assembly/Homologous Model

3.1.1.1 Mobility of a Spool/Equilibrium at a Working Point

At a constant command signal, the servo valve establishes generally an equilibrium state at the actual working point where all internal leakages are accordingly set. At such a balanced state, the spool perseveres in its position in the leakage stream. This can be compared with a (freely) hovering ping-pong ball in a vertical air stream: whenever the resulting force of the flow is reduced, the ball will sink down. And it will rise when the flow force increases. Whilst hovering, the exchange between the potential energy and the resulting kinetic energy will be balanced anew at every moment. Note that the associating forces are varying continuously because the equilibrium at a working point has to be continuously re-established. As a result, the ball keeps on hovering in a certain, more or less 'constant' altitude level even though it carries out continuously a chaotic movement in all directions. This thought experiment is very useful to complete the measurement data interpretations discussed in Sect. 2.1.2.

3.1.1.2 Confinement of the Spool's Freedom Grade/Guided Sliding of the Spool

In Sect. 2.1.2.2, it was discussed that there is a significant coherency between the mobility of the spool and the stability of the pressure signal answer. The stability of the signal answer can be considered now by extending the thought experiment discussed in the previous chapter; in the next step, the hovering ball shall be equipped with a ballast weight as shown in Fig. 3.1. Its vertical movement implies directly the longitudinal position of the spool and consequently represents the pressure level at the outlet of a valve.

Placing cylinders of different diameters vertically onto the ball, the degree of freedom of the ball can be limited; in a tightly fitted cylinder, the ball is able to move only in a vertical direction (cf. the L/H scheme in Fig. 3.1). The greater the gap distance given between the cylinder wall and the ball, the less is the limitation of movement in the actual horizontal plane perpendicular to the longitudinal axis (cf. the middle scheme in Fig. 3.1). Whenever the cylinder is misaligned to the ball with an extra angle as shown in the R/H scheme of Fig. 3.1, the ball possibly leans on the cylinder at a certain position. In the worst case, the ball cannot move in the cylinder at all. At a lesser misalignment, the ball will come into contact with the cylinder wall and it will be guided in the longitudinal direction of the cylinder. As long as the ball is still able to move, the valve system does not experience a total blockage. However, this contact will result in quasi-sedation in the plot of the pressure history due to the leaning of the spool against the sleeve wall (Note that the final result will be a sluggishly moving spool caused by the significantly increased friction. This step will be discussed in the next chapter. In this chapter, the possible increase of fiction will be not considered yet).

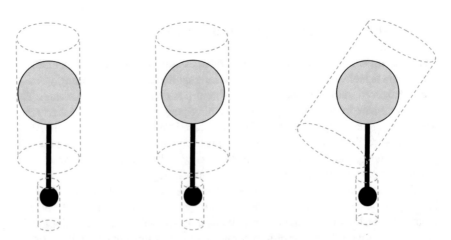

Fig. 3.1 Geometric tolerances and the spool's freedom grade

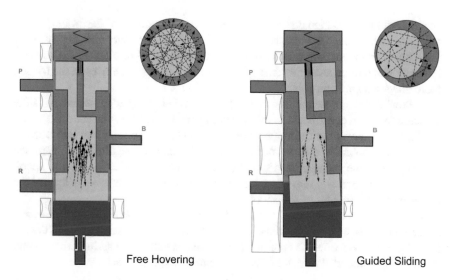

Fig. 3.2 Confinement of the spool's freedom grade/guided sliding of the spool

Considering these facts, the plots presented in Sect. 2.1.2.2 can now be easily understood. The hyper-stable section in the plot is nothing but a reflection of a direct contact of the spool to the sleeve (cf. Figs. 2.3, 2.4, 2.5, 2.6, 2.7, 2.8, 2.9, 2.10 and 2.11).

A hyper-stable signal answer, however, does not mean that the system is already faulty or going to be implicitly so, but it is rather a sign that the system is on the 'threshold' and vulnerable at the moment. Note that the spool recovers occasionally such a situation as shown in Figs. 2.3, 2.4 and/or 2.6. In order to understand the fault mechanism, it is of importance to know what such a marginal case looks like in reality. Figure 3.2 shows the difference of both situations schematically (Note that the different numbers of droplet symbol depict qualitatively the actual intensity of the leakage at each 2D-projected location.). In the normal case, the hovering spool carries out a chaotic movement in all directions. The leakage is consequently in the nominal or design value and keeps its average level (cf. the L/H scheme in Fig. 3.2). When the spool leans on the wall, the movement is no longer so chaotic. The pressure variation at the brake port reflects this even though the mean pressure keeps its level more or less unchanged (cf. Figs. 2.3, 2.5, 2.7 and/or 2.9).

The conclusion is that the hyper-stable signal answer is nothing but a reflection of a guided sliding movement of the spool on the sleeve's wall.

3.1.1.3   Equilibrium and Meta-Equilibrium in the Second Stage

The reason for the fault traditionally called 'hydraulic locking' is a temporarily increased internal friction inside the spool and sleeve assembly. Generally, the

actual friction can change due to the non-perfectly manufactured hardware. This is
the case, when the non-ideally round and straight borehole and the likewise
non-perfect spool are coincidently misaligned due to actual insufficient hydro-static
lubrication condition and/or thermal effects (cf. Sect. 3.1.3). In any case, the friction
force plays a significant role.

Following the considerations made in the previous chapter, a boundary faulty
situation will be reconstructed in this chapter by analysis of force balance at a spool.

Figure 3.3 shows the definition of relevant forces and pressures around a spool.

When the spool is still properly working in spite of significantly increased
friction, the force balance is depending on the running direction of the spool,
whereas the friction is an inversed signum function and accrues always against the
actual running direction of the spool.

The upper scheme in Fig. 3.4 shows the difference of force balances in accordance with the actual running directions. In the case of pressurizing, the hydraulic
force F_H must overcome the resulting force sum consists of the spring force F_S and
the resisting force of the friction F_{F1}. This can be expressed as:

$$F_H > F_S + F_{F1} \tag{3.1}$$

In contrast, in order to reset the system, i.e. to bring the spool back to the starting
point, the spring force F_S must be strong enough to overcome the sum of the
hydraulic force F_H and the actual friction force F_{F2}. Then, the inequation has to be:

$$F_H + F_{F2} < F_S \tag{3.2}$$

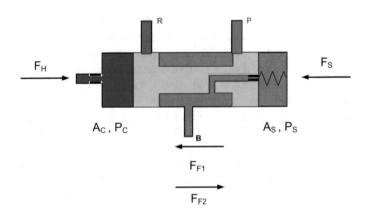

Hydraulic Force : $F_H = Abs(A_C \times P_C - A_S \times P_S)$
Friction Force: F_{F1} : whilst pressurizing
                          F_{F2} : whilst releasing
Spring Force: F_S

Fig. 3.3 Definition—forces around a spool

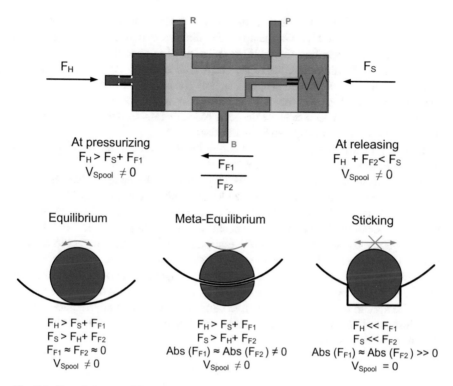

Fig. 3.4 Force balance—differences between equilibrium and meta-equilibrium

Note that the hydraulic force F_H is only the hydraulic resistance caused by the fluid dumping process in this case. This is valid until reaching the next lower working point.

In normal case, the friction force plays only a subordinated role in the force balances. Should the friction force increase due to any reasons, the spool becomes sluggish. So far the spool is still manageable, the state can be designated as 'stable'. The system will try to compensate the additional force at any time regardless the working point and the running speed of the spool. This is also a kind of equilibrium. However, it differs from the equilibrium discussed in Sects. 3.1.1.1 and 3.1.1.2 as the friction force plays now a significant role besides the hydro-dynamical flow forces. In order to distinguish this state from the equilibrium of hydro-mass balancing, the phenomenon of hydro-mechanical balancing will be named hereafter as 'meta-equilibrium'. At any time in both equilibrium and meta-equilibrium cases, the spring force F_S is perfectly balanced by the sum of the counter forces given by the hydraulic and friction forces.

When the friction force is increased furthermore and eventually high enough due to any reasons, the spool comes to standstill (sticking).

It must be kept in mind that both equilibrium and meta-equilibrium are quasi-static states. The demanding speed of the spool is negligible in both situations, even if the spool still carries out a chaotic movement by itself. The symbolic schemes in Fig. 3.4 illustrate the difference between equilibrium and meta-equilibrium regarding the mobility/freedom of the spool; in the case of the equilibrium, the freedom/mobility of the ball and accordingly the chaotic movement discussed in the previous chapter is not restricted. In the case of the meta-equilibrium, however, the ball can move only with a restricted freedom grade and is not necessarily able to leave the area. Nevertheless, the situation differs from a total blockage (sticking). Figure 3.4 illustrates this.

Figure 3.5 clarifies the reflection of the difference between the equilibrium and meta-equilibrium in the pressure signal answers.

According to the considerations discussed above, the sudden changes in the dithering of the curve run shown in the plot are nothing but a certain 'equilibrium mode changing' inside the spool and sleeve assembly. The reason for a recovery of the spool's mobility can be of a wide spectrum of physical phenomena, like from a simple agitation/vibration to nonlinear complex pulse modulation caused by different hydraulic circuit members or even the secondary effect of their mutual interferences. Accordingly, the countermeasure against the meta-equilibrium could also have a wide spectrum of possibilities. In any case, the most efficient way to make the system robust is the preventive elimination of the possible sources of the fault instead of amendments, which could occasionally work but they reduce only the intensity of commenced fault.

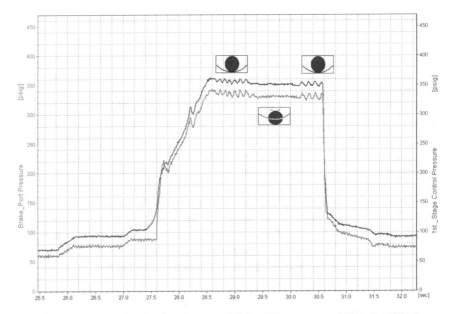

Fig. 3.5 Differences in the signal response; equilibrium versus meta-equilibrium at GRB 2

There is no doubt that the hyper-stable signal response shown in Figs. 2.3, 2.4, 2.5, 2.6, 2.7, 2.8, 2.9, 2.10 and 2.11 is a reflection of a sudden, temporarily increase of the friction. It is expected that the spool's movement is no longer so chaotic, especially in longitudinal direction since the friction suddenly plays a significant role and only in this direction. In Figs. 2.3, 2.4, 2.5, 2.6, 2.7, 2.8, 2.9, 2.10 and 2.11, it is easily to recognize that the mean pressure level is kept more or less constant. The only changing in the curve runs is the amplitude of the variation whilst the spool tries to establish equilibrium at every moment. Nevertheless, these do not necessarily evidence enough that the spool really got in touch with the sleeve's wall and was guided by it.

Figure 3.6 confirms representatively such a contact. A frequency analysis showed that the signal response at $t < 29.375$ and $t > 31.165$ is a superposition of two major frequency bands. In contrast to the high frequency of approx. 18.9 Hz, which changes the intensity only slightly throughout the whole time range, the lower frequency of approx. 8.93 Hz damped down to an approx. 0.67 Hz level in the time $29.375 \leq t \leq 31.165$ and restarted after passing $t = 31.165$. The higher frequency is identified as the dithering frequency of the spool perpendicular to the spool's longitudinal axis. Considering the mass moment of inertia regarding the reference axis, the lower frequency must be that of the spool's movement in the longitudinal direction.

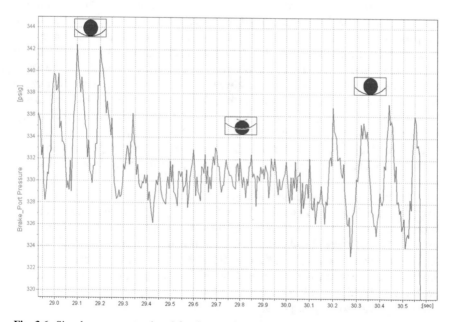

Fig. 3.6 Signal response at reduced freedom grade at the spool

It is concluded that the sticking is a result of a certain reduction of freedom grad of the spool caused by increased friction. Although it does not necessarily mean a direct contact of the spool to the sleeve, the consequent non-coaxial alignment of both members increases implicitly the local gap distance and eventually the internal leakage (cf. Fig. 3.2).

3.1.2   Origin of the Fault/Interlock Mechanism

3.1.2.1   Order of the Fault Induction/Origin of the Fault

According to the observations, the two-stage servo valve did not react at all in some serious cases even at the electric connector removal, i.e. the valve experienced a real 'hydraulic locking'. In some less serious cases, the measurement results showed that the pressure answer was no longer exactly to be correlated with the command signal. In other events, there were reactions which do not reflect the command signal whilst dropouts are registered in the second stage.

This, however, does not yet explain the origin of the fault.

The fact is that the first stage is able to shift the spool by means of hydraulic pressure only forward in longitudinal direction against the return spring. And the spool should return to the start position as soon as the pressure in the control line is released. Aside from blockages caused by debris, which is not to be considered for the present fault symptoms (cf. Sect. 1.2.3.1), the only possible significant reason remaining for a total blockage is of a fluid dynamic nature, as long as sufficient geometric tolerances are chosen (cf. Section "Dimensional Tolerances at the Spool and Sleeve Assembly"). The friction between the spool and sleeve does not have to be high enough to keep the spool in position alone. Not even the 'traditional' sticking requires this (cf. Section "Changing of Actual Friction on the Sliding Surface Area Whilst Operating" and Sect. 3.1.1.3).

Disregarding such factors/influences, which are unlikely to play a role, there remains only the possibility of equilibrium to explain the phenomenon in the first stage.

When a constant flow is established at the control line and the state is frozen, a part of the kinetic energy in the flow keeps the spool in a certain position against the return spring as well as the same thing happening around the flapper at the same moment. Then, on both locations certain equilibrium has been established between the local mechanical forces[1] and the local flow force. The very basic prerequisite for this situation, however, is a sufficient flow rate in the control line. Such flow rate can be established only when the internal leakage in the spool and sleeve assembly has been previously increased due to the necessary drain rate. This is not possible

---

[1]i.e. spring force and friction force, whereas no friction force exists in the first stage at all.

for a properly working spool and sleeve assembly unless the spool falls somehow into a misalignment relative to the sleeve axis (cf. Section "Internal Leakage and Gap Tolerance as a Significant Influence Factor").

In summary, the abnormality in the first stage can only be initialized and consolidated by the spool's misalignment in the second stage. Hence, the origin of the fault investigated in the present troubleshooting must be the second stage (cf. Sects. 3.1.2.3, 3.1.3.1 and 3.1.3.2).

3.1.2.2 Self-stabilization of the Flapper/Interlock of the Entire Servo Valve

The first stage can still manage the spool up to a certain grade if the flow condition once leaves laminar flow range and falls only slightly into the sub-critical flow range as the flapper sporadically experiences there relative weak suction force due to the instable separation of the flow from the flapper contour (cf. Sect. 1.2.5). Passing this range, i.e. falling deeply into higher Reynolds number region, the first stage almost loses its controllability due to constant and stronger suction force. After passing that sub-critical flow range, the strong jet stream from the inlet nozzle starts walloping the flapper further away and keeping it in position. When eventually the system falls into this super-critical flow range, where the transition point on the cylindrical flapper body wanders further downstream, the first stage stabilizes by itself. The result is that the flapper is no longer manageable at all. Although the strength of the jet stream accordingly sinks down if any disturbances possibly reduce the drain, the flapper is hardly able to cross the sub-critical flow range due to the suction force. Hence, the flapper position stabilizes by itself in a point behind the full-open position of the normal operation case. This report does not go into detail on this complex phenomenon whereas the pressure condition and the geometrical shape, like the form and dimensions of the effective cross section, actual gap geometry at the nozzle, are changing and accordingly the angle of incidence changes in a nonlinear manner. In any case, these rebound effects of the interplay given by the shuttling between two effective Re-Number regions stabilize/consolidate the equilibrium state with a certain reserve damping margin against possible disturbances. This will result in an interlock of the entire servo vale.

3.1.2.3 Danger of the Flow Separation on the Flapper Body and the Timing

The flow around the flapper body separates from the cylindrical shape as soon as the actual Re-Number exceeds the critical range. Considering the fact that the geometric dimension is unchangeable and the changing of the medium's kinematic viscosity is negligible for a short interval, the local Re-Number depends solely on the actual velocity of the flow. As shown in Fig. 3.7, such sporadically occurring transit phase is clearly to identify in the plot; the curve run features an interim,

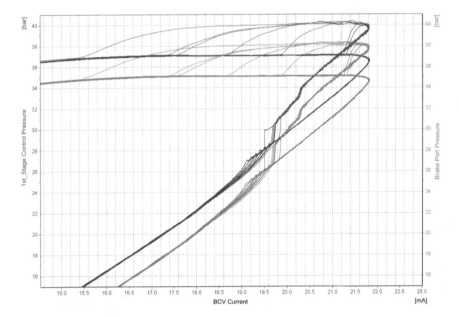

Fig. 3.7 Sporadic shifting of signal response

continuously changing gradient between the 'nominal track' and the parallel upper shifted one. At the end of the transit phase, the curve runs into the normal track.

Such a track changing is also found in Fig. 1.4 in the low current range of 13.5–20.5 mA.

This phenomenon occurs whenever the actual *Re*-Number crosses over the critical point of *Re* = 2330. Note that the flow rate and accordingly the actual velocity of the fluid is depending on a large number of linear and nonlinear parameters, like dynamic viscosity of the system fluid, gas content in the fluid, temperature, natural frequency of the system, pressure variation and even on the acoustic wave during the operation.

The first and second stages do not automatically escalate the situation, when the spool reacts sluggishly due to the increased friction (i.e. reduced reactivity) or the flow separates from the cylindrical flapper. As discussed in the previous chapter, the prerequisite for the interlock working mechanism is a high internal leakage flow. If the flapper is still manageable by the control electronic, the flapper can shut off the inlet nozzle.

However, the misfortune is perfect, if the increased internal leakage due to the misalignment at the second stage and the flow separation at the flapper body happen coincidentally and simultaneously.

The transit shown in Fig. 3.7 itself is harmless as long as the spool and sleeve assembly works properly. As shown in the plot, both stages work synchronized

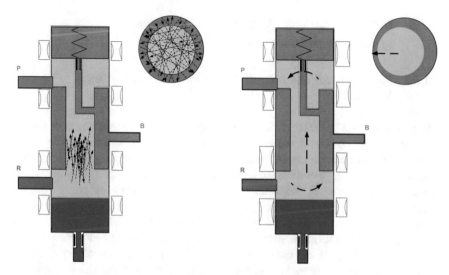

Fig. 3.8 Transit phase and occurring of a coaxial misalignment in the second stage

despite shifted pressure behaviours. Note that the servo valve was demanded by a constant demanding speed. Solely the releasing speed was varied; cf. Fig. 1.17. The transit phase, however, is the most dangerous phase because the first stage is exposed to the danger of 'freezing' without any recovery chance. At the same time, the second stage is also exposed to the danger of 'non-coaxial misalignment' if the relief grooves are not working properly. Figure 3.8 explains the circumstance schematically (Note that the different sizes of restrictor symbol depict qualitatively the actual intensity of the leakage at each 2D-projected location.).

Due to the dominant longitudinal movement, the spool carries out no longer a chaotic radial movement during the transit phase. The arising danger is that the spool is exposed to a lateral force without having a chance to coaxial realignment. Hence, the spool is endangered easily to incline and then if it occurs, it leans very possibly on the sleeve's wall. If such non-coaxial misalignment of the spool occurs during the transit phase, the flow around the flapper in the first stage will be increased as a secondary effect. The self-induction is completed, and the loop is closed in this way. Closing the self-induction loop, the Reynolds number falls deeper into the critical region, where the first stage has no chance to recover the situation due to the self-stabilization discussed in the previous chapter. In contrast, the spool occasionally recovers the non-coaxial misalignment alone (this is also evidence that the second stage is the origin of the fault; cf. Sect. 3.1.2.1).

The conclusion is that the system is particularly endangered when the flow separates from the flapper body, even though the first stage is not the origin of the fault (cf. Sects. 3.1.2.1, 3.1.3.1 and 3.1.3.2).

3.1.3   Initializing, Propagation and Escalation
          of the Blemish/Reason for Blemish

3.1.3.1   Degradation/Collapsing of the Lubrication Film—'Dry Friction'

The plot shown in Fig. 3.9 (Fig. 2.2) is a history of serial command cycles before the fault has finally been registered as such. It is striking that the working ability of the valve system became worse and worse step by step. Considering that the performance of the first stage can hardly be changed stepwise but rather suddenly and in the manner of 'either/or', the behaviour of the signal answer must come from the mechanical moving part of the second stage, which can certainly be stepwise and more sluggish whenever the working condition of the sliding surface gets worse. It is an indication that the lubrication film condition has a significant effect on the self-induced locking. When the process is ongoing, the lubrication ability of the oil film seems to rapidly degenerate within two or three cycles. Once the degeneration of the lubrication has commenced, the mobility of the spool is limited and finally the spool comes to a standstill. The spool occasionally frees by itself so that it works properly again. This happens dramatically if the spool recovers its mobility, for instance as a result of vibration and/or shaking (cf. Fig. 3.7). Even this signifies that the self-induced fault is initialized by a pure mechanical problem. Whenever two metallic surfaces start rubbing into each other due to a poor lubrication, the thickness of the film must be reduced and consequently the gap between two circular parts

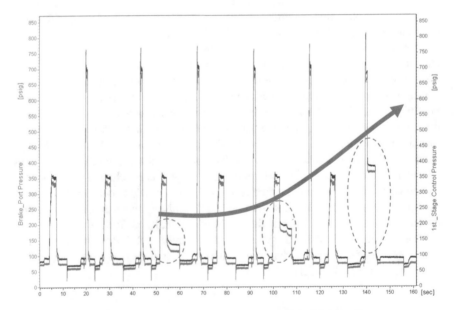

Fig. 3.9 Progressive fault intensity due to the degeneration of the lubricant film

completely disappears or at least is significantly reduced. In the case of cylindrical slide sudden high internal leakage is expected as the consequent result.

Hence, a sluggish mobility of the spool is a herald of the self-induced hydraulic locking.

The control loop of the brake system recognizes an abnormality as a fault only when the actual parameters of dwell and pressure level exceed the preset threshold values (cf. Sect. 2.1.2.1). Hence, it would be purposive to distinguish a 'complex fault' from the 'simple fault', whereas the latter should be representative of the classic 'hydraulic locking' or temporal jamming whilst the former features the self-induced fault discussed in this report (cf. Fig. 3.13). Note that this is only a terminological differentiation as the simple fault is in this special working mechanism of the self-induced fault only an 'unripe fault' or 'incomplete fault' from the system engineering point of view.

3.1.3.2   Holistic Consideration Incl. The Reaction of the Control Loop

The starting point of the blemish is uncertain and not easy to predict (cf. Sect. 3.1.4). It is also hardly possible to exactly determine the actual friction between the spool and the sleeve. Nevertheless, the starting point of the blemish is easy to recognize if the spool is already in motion; when the spool became suddenly sluggish, the actual friction had been increased or both parts in the second stage might have come into contact (cf. Sects. 2.1.2.2 and 3.1.1.2).

Having no reference to be compared, however, the abnormality is not easy to recognize. Figure 2.9, for example, shows on its right-hand side such an interesting case. Although this is the worst case in the plots shown in this investigation, the behaviour of the pressure from both stages looks inconspicuous.

But, in truth, the friction changing in the second stage has occurred before the servo valve is demanded. This can be concluded by analysing the electric power consumption in the commencing phase of the demanding. Compared to the current peaks at the coil during normal operations (approx. 23.3 mA, see Figs. 2.8 and 2.11), that in the fault case is significantly higher (24.1 mA, see Fig. 2.9) for a similar pressure answer of ca 750 psi.

This means that the control of the servo valve had tried to overcome/compensate the high friction in the second stage by demanding the coil with a higher current. Note that the control system works in closed loop considering the actual pressure answer as a feedback signal.

Hence, it is concluded that the first stage is not affected by any means at the moment of the demanding, and consequently the induction based on mutual interference has not commenced at that time [by the way, this is also direct evidence again that the second stage is the origin of the fault (cf. Sect. 3.1.2)]. The second pressure peak reflexes the dynamical spring-back of the flapper in motion. In normal case, the control system can manage the reaction by reducing the current so that the second peak cannot be too high. In the case of the fault, it is recognized that the reaction of the command (current) has a certain phase shifting. And the system falls into a certain overreaction despite closed control loop. The reason for such overreaction with a phase shifting is the unidirectional aiding force at the flapper. This comes from increased flow rate on the outlet in the first stage (cf. Sects. 3.1.2.2 and 3.1.2.3).

After this event, the control loop seems to have completely lost its hegemony. The first stage is no longer manageable. The plot in Fig. 2.9 shows this: passing $t = 139.75$ s, the pressure signals do not react at all despite a more than 2 mA changing in the current level of the command. According to the plot, there was still pressure in the command line between the first and the second stages after the shutdown of the current at $t = 140.43$ s. This is surely caused not by obstruction but by an internal flow which keeps the flapper in opened position.

Figure 2.10 shows an interesting intermediate case. The control electronic set the current at the coil to approx. 24 mA, presumably due to the increased internal friction at the spool and sleeve. During the commencing phase of the demanding, the spool became free to move thanks to some pressure variations, i.e. the dithering at the pressure development recovered the spool's movement and consequently the pressure answer did not significantly deviate from the set value. The details of the Figs. 2.9 and 2.10 are shown in Figs. 3.10 and 3.11, respectively. It is easy to recognize that the commencing phase of the normal case, i.e. Figure 3.11/PPT 1, contains at least four pressure variations.

The plots discussed in this chapter might be considered as a confirmation of the working mechanism reconstructed in previous chapters (Fig. 3.12).

3.1.4 Arbitrariness of the Fault—Russian Roulette Effect

The arising question is now why the self-induced hydraulic locking happens randomly so that the fault is not always to be observed in spite of unchanged working conditions. In this chapter, the reason for the random scattering will be discussed.

The fault potential described in Section "Dimensional Tolerances at the Spool and Sleeve Assembly" very much resembles 'Russian roulette' as the actual angular position of a rolling, non-perfect cylindrical body in a likewise non-perfect borehole will be randomly determined by hydraulic flow which passes the cross section of the assembly. Consequently, its amount is hardly predictable. Furthermore, during operation, a quasi-axial stream will be determined by unpredictable internal leakage at a given coincidental position. The stream in axial direction can be either aiding or inhibiting. This additional fact makes the prediction of spool's actual angular position fully impossible.

Even though the susceptibility of an angular section caused by the misfortunately aligned parts could be identified in advance, the fault occurs arbitrarily as long as the inevitable geometric misalignment inclines to play a dominant role (cf. Section "Importance of the Balancing Grooves for Spool's Working Ability").

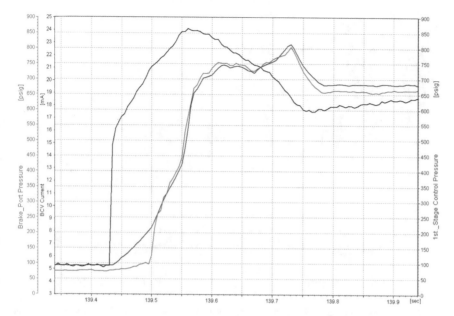

Fig. 3.10 No dithering during the commencing phase of PPT No. 6; abnormal/fault

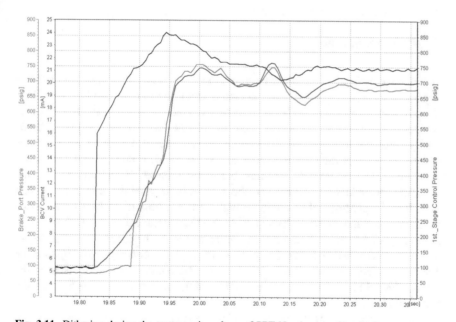

Fig. 3.11 Dithering during the commencing phase of PPT No. 1; normal/no fault

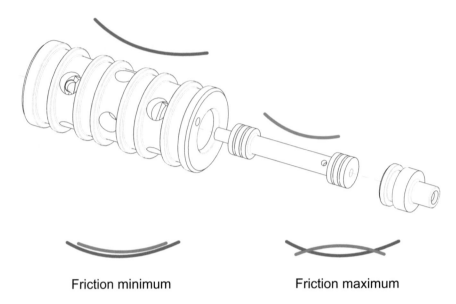

Friction minimum   Friction maximum

Fig. 3.12 Reason for randomly changing friction in the spool and sleeve assembly—Russian roulette effect

3.2   Conclusion and Summary

3.2.1   *Conclusion*

A two-stage hydraulic control valve consisting of an electromagnetic servo valve of flapper type and a spool and sleeve assembly can lose its controllability completely if both stages influence each other mutually. Becoming sluggish due to direct contact of the spool to the sleeve wall, the second stage initializes the fault and finally escalates the situation by starting the process of a high internal leakage caused by misalignment of the spool in the sleeve. The only reason for such a misalignment is improperly working balancing grooves in the sub-component. Once the self-induction commences, the process is irreversible as soon as the first stage loses its controllability due to the high flow rate. Should the flow force caused by internal leakage not be high enough, so that the electric coil and the return spring of the flapper are still able to manage the movement at the spool by themselves or they even overcome the blockage by some chance, e.g. vibration or jolting, then the system recovers its controllability. Whenever the control loop still manages the servo valve within the predefined threshold of timeout, the fault can be masked and remain undetected.

The reconstruction of the entire working mechanism is shown in Fig. 3.13 as a flow chart of the fault scenarios.

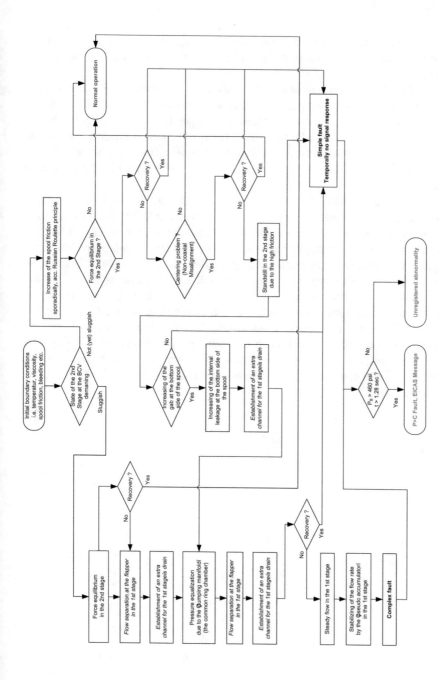

Fig. 3.13 Flow chart for fault scenarios with predefined threshold values of timeout and pressure level

3.2.2 Summary

Two-stage hydraulic servo valves equipped with a spool and sleeve assembly and a nozzle and flapper type first stage can fall in a total blockage due to fluid dynamic phenomena.

The initialization happens at the second stage in the manner of 'Russian roulette' principle. Moving in a non-perfectly straight and non-ideally round borehole, the likewise non-perfect spool can coincidently misalign itself due to actual insufficient hydro-static lubrication condition if the balancing grooves in the spool and sleeve assembly are not working properly. As soon as the spool stands no longer coaxially, the internal leakage increases exponentially due to the viscous hydraulic fluid as working medium.

After being initialized by such a mechanical misfortune, the internal leakage flow can develop between the spool and the sleeve so critically that the flapper in the first stage can barely manage the inlet nozzle despite the support of the pre-loaded spring force. In the worst case, the flow separates behind the cylindrical flapper body. Due to the increased drag, the force balance is so seriously disturbed that the flapper drifts away from its position as if a sudden extra opening force has been developed. Then, the servo valve is eventually no longer manageable by electric command signals. Once induced by the second stage to the fault, the first stage stabilizes the state against disturbances since the flow condition around the cylindrical flapper varies/shuttles between sub-critical and super-critical Reynolds number regions. Despite all these interactive fluid dynamic phenomena, it must be said that the root cause is improperly working balancing grooves in the spool and sleeve assembly.

Epilog

The troubleshooting was a quite exhausting task in all respects.

Besides technical and scientific challenges, there was a lack of analytical mindset among the task force members, too.

The author is so bold as to tell a story with those colleagues from the involved partner companies in mind:

Once upon a time a schoolboy felt largely bored in the summer vacation. He decided to investigate the learning attitude of a grasshopper. Soon he caught a magnificent exemplar in the garden of his parents' house and started training it. He placed the magnificent guy in the middle of the table. Then, thumping the table, he shouted "Jump". Almost every time the grasshopper jumped on the table. The boy recorded his survey in writing:

"Day 1: It is not so easy to train a grasshopper but I was able to instill obeying into him. On my command my grasshopper jumps every time. I am proud of my teaching performance......"

A few days later, however, the boy found it dull to repeat only the same. And the grasshopper seemed not to be able to learn anything else, neither dancing nor somersault, let alone singing. But, the boy wanted to make a detailed study and ask himself a 'scientific' question; "what happens when I remove one of the anklebones of the grasshopper?" No sooner thought than done. He performed an amputation. One of the grasshopper's anklebones was removed and the boy tried the same training. Thumping the table, he shouted "Jump" to the grasshopper. The grasshopper jumped up again and again whenever the boy thumped the table. The boy reported the result in his notebook:

"Day 5: One leg amputated. Nevertheless, on my command my grasshopper jumps every time. There is no significant changing in his behavior......"

A few days later the boy found it dull again to repeat only the very same. As next step of his study he decided to remove the other anklebone as well. No sooner had he amputated the leg than he started the experiment. Again, thumping the table, he shouted "Jump" to the humble grasshopper. The grasshopper, however, jumped no longer this time. The boy reported the result in his notebook as usual:

"Day 8: Both anklebones amputated. The grasshopper jumps no longer. Having had both anklebones amputated, grasshoppers become deaf."

continued to the next page

T. Seung, *Self-Induced Fault of a Hydraulic Servo Valve*,
SpringerBriefs in Applied Sciences and Technology,
https://doi.org/10.1007/978-3-030-03523-5

Dear reader,

It's funny, isn't it?
Were you laughing at the boy?

May I ask you, why? ....?!

Did you know that a large part of grasshoppers have their acoustic hearing organs around their knees of the anklebone?

The moral of the story:
We often do conclude our cognition prematurely without knowing
much about our subject. Things look like logical and seem to work
sometimes, even if we made incorrect conclusions and wrong decisions.
Also, things looked wrong, even if we made correct conclusions and
right decisions. The scientists are trying to find the truths in order not to
fail next time.

Taenny

Amendment — Never mind?

For those who do not feel be beaten or insulted:

Some of you seemed to capitulate as soon as you heard that a large part of grasshoppers have their acoustic hearing organs around their knees of the anklebone.

I wonder why!
It was not to experience whether the grasshopper of the boy was one of such sort or not.

We occasionally give up too quickly and start immediately coming to arrangement with unknown and undefined subjects, when we somehow feel as if we had been caught. We even let us become intimidated.
We —scientists and engineers— ought not to give up so easily.
Don't you think so?

Very sincerely

T. Seung

References

1. Acohido, B.: Pittsburgh disaster adds to 737 doubts. The Seattle Times (October 29, 1996), see also "The 1997 Pulitzer Prize Winners Beat Reporting". http://www.pulitzer.org/archives/5926
2. Backé, W., Tatar, H: Untersuchung des Einflusses von Störkräften auf den Schaltvorgang bei Wegeventilen in der Hydraulik. Westdeutscher Verlag (1975)
3. Blackburn, J.F.: Contributions to hydraulic control—lateral forces on hydraulic pistons. Trans. ASME **75**, 1175–1180 (1953)
4. Borghi, M.: Hydraulic locking in spool-type valves: tapered clearances analysis. Proc. Inst. Mech. Eng. **215**, 157–167 (2001)
5. Dransfield, P., Bruce, D.M., Wadsworth, M.: A general approach to hydraulic lock. Proc. Inst. Mech. Eng. **182**, 595–602 (1967)
6. Lohm Laws, N.N.: Chapter M / Technical Hydraulic Handbook Eleventh Edition 2009, The LEE Company, Connecitcut / USA
7. Manhajm, J., Sweeney, D.C.: An investigation of hydraulic lock. Proc. Inst. Mech. Eng. **169**, 865–602 (1955)
8. Milani, M.: Designing hydraulic locking balancing grooves, Proc. Inst. Mech. Eng. Part I: J. Syst. Control Eng. **215**(5), 453–465
9. Schlemmer, K.: Klemmkräfte an Schieberwegeventilen - Analyse der Querkraftreduktion durch Druck ausgleichende Umfangsnuten. *Zeitschrift für Fluidtechnik Aktuatorik, Steuerelektronik und Sensorik*49. Jahrgang Heft 7. S. 443–447 (2005)
10. Schlemmer K.: Druckentlastungsnuten an hydraulischen Ventilschiebern. http://www.ifas.rwth-aachen.de/Main/Forschung/projekte/sl.html
11. Schlemmer, K.: Reduction of hydraulic locking forces in spool valves by flow analysis of grooved small gaps. http://www.ifas.rwth-achen.de/Main/Institut/IFK%20Paper/D3_schlemmer_paper.pdf
12. Schlemmer, K., Murrenhoff, H.: Auslegungsmethodik zur Druckentlastung von Ventilkolben. Automatisierte Simulation als Konstruktionshilfe. Zeitschrift für Fluidtechnik Aktuatorik, Steuerelektronik und Sensorik **50**, Jahrgang Heft 7. 10–19 (2007)
13. Servo Valve with Mechanical Feedback D761 Series ISO 10372 Size 04. moog.com/literature/ICD/Products/Valves/d760seriesvalves.pdf
14. Sweeney, D.C.: Preliminary investigation of hydraulic lock. Engineering **172**, 513–516, 580–582 (1951)
15. Sweeney, D.C.: Eight ways to overcome hydraulic lock. Engineering **190**, 592–593 (1960)
16. Whiteman, K.J.: Hydraulic lock at high pressures. The British Hydromechanics Research Association Publication. RR 521 (1955)

© The Author(s), under exclusive license to Springer Nature Switzerland AG 2019
T. Seung, *Self-Induced Fault of a Hydraulic Servo Valve*,
SpringerBriefs in Applied Sciences and Technology,
https://doi.org/10.1007/978-3-030-03523-5

Printed in the United States
By Bookmasters